T0122977

Developing a Risk Assessment Methodology for the National Aeronautics and Space Administration

Daniel M. Gerstein, James G. Kallimani,
Lauren A. Mayer, Leila Meshkat,
Jan Osburg, Paul Davis, Blake Cignarella,
Clifford A. Grammich

Prepared for the National Aeronautics and Space Administration

For more information on this publication, visit www.rand.org/t/RR1537

Library of Congress Cataloging-in-Publication Data is available for this publication.
ISBN: 978-0-8330-9563-3

Published by the RAND Corporation, Santa Monica, Calif.
© Copyright 2016 RAND Corporation
RAND® is a registered trademark.

Cover image: NASA. CubeSats deployed from the International Space Station, May 17, 2016

Cover design by Eileen Delson La Russo

www.rand.org

Preface

The Study

This study was conducted in response to a request from the National Aeronautics and Space Administration (NASA) to assist in gaining an understanding of how various disparate factors contribute to overall risk associated with NASA-level decisions. The specific mandate was to "provide NASA's Office of Strategy and Plans with one or more risk frameworks that integrate risk factors and risk management approaches tailored to NASA's management, operations, and acquisition structures."

NASA confronts a variety of organizational-level risks within its programs. Comparing, contrasting, and mitigating these risks require developing a common lens through which to view the risks. Such an evaluation can provide opportunity to gain an overall understanding of the risks associated with NASA-level decisions. This study developed such a methodology. Specifically, it resulted in the development of a single risk-informed decision support and includes the identification of various techniques for conducting risk assessments and the definition of the seven risk factors identified as being important for NASA-level decision consideration.

The report also includes identification of the risk factors, components, indicators, and mitigation strategies for two case studies: (1) cancellation of the Space Shuttle program and commercialization of transport to the International Space Station and (2) the Cislunar Habitat.

This report is written for a non-expert audience so that any practitioner or decisionmaker, with any level of training, could make use of the methodology. It therefore does not assume knowledge about risk or decision analysis on the part of the reader.

RAND Science, Technology, and Policy

The research reported here was conducted in the RAND Science, Technology, and Policy program, which focuses primarily on the role of scientific development and technological innovation in human behavior, global and regional decisionmaking as it relates to science and technology, and the concurrent effects that science and technology have on policy analysis and policy choices. The program covers such topics as space exploration, information and telecommunication technologies, and nano- and biotechnologies. Program research is supported by government agencies, foundations, and the private sector.

This program is part of RAND Justice, Infrastructure, and Environment, a division of the RAND Corporation dedicated to improving policy- and decisionmaking in a wide range of policy domains, including civil and criminal justice, infrastructure protection and homeland security, transportation and energy policy, and environmental and natural resource policy.

Questions or comments about this report should be sent to the project leader, Daniel Gerstein (Daniel_Gerstein@rand.org). For more information about RAND Science, Technology, and Policy, see www.rand.org/jie/stp or contact the director at stp@rand.org.

Contents

Figures and Tables

Figures

Tables

Summary

Background and Purpose

The National Aeronautics and Space Administration (NASA) confronts a variety of organizational-level risks within its programs that could influence the success of its missions or programs. Comparing, contrasting, and mitigating these risks require developing a common lens through which to view them. Such an evaluation can increase overall understanding of the risks associated with NASA-level decisions.

This report provides NASA's Office of Strategy and Plans with a risk assessment methodology that integrates the factors of risk with management approaches tailored to NASA's management, operations, and acquisition structures. While NASA has deep experience in conducting risk assessments on highly technical issues for individual programs and projects, it has not developed a methodology for thinking about overall risk associated with its decisions.

Developing one overarching risk-informed decision support methodology was the goal of this effort. We present such a methodology for a non-expert audience so that any practitioner or decisionmaker with any level of training can use it. We do not assume knowledge about risk or decision analysis on the part of the reader.

Study Approach

Our multidisciplinary effort entailed a review of existing literature on risk assessment and management, identification of and definitions for overarching risk factors (i.e., those factors that may influence NASA's mission success), development of an informed decision methodology, and validation of this methodology through two case studies.

In conducting this study, we focused on the risk factors, best practices, and associated risk management approaches relevant to NASA's management, operations, and acquisition. The intent was to inform NASA's ultimate interest in evaluating specific risk postures of the agency; such evaluations are beyond the scope of this study but could be follow-on projects for RAND.

We leveraged RAND expertise across the applicable risk domains and consulted with subject-matter experts to search for other relevant approaches. We then analyzed those approaches and integrated them into a multitiered, cross-functional, risk-informed decision methodology for use in evaluating risk posture across various dimensions and levels of NASA.

Any organization faces a range of risks to meeting its objectives. Risks cannot be eliminated, but they can be consciously managed and traded through various techniques that prioritize investments and policies. This study examines the range of risk factors and existing approaches for managing those risks.

The risk factors that we consider are supply chains; acquisition costs and schedules; human capital or workforce risk (both civil servants and support contractors); organizational and managerial considerations (e.g., issues arising from NASA's distributed management between mission directorates and research centers); external dependencies; domestic and international political risks (e.g., partnership instabilities; congressional equities) and technical issues, including information security.

Developing the Methodology

We identify a methodology for consolidating disparate risk factor inputs—gathered either from existing risk assessments in use by NASA or by conducting alternative assessment techniques for factors where no risk assessments currently exist—to develop a list of risks across the seven factors listed above.

In developing the methodology, we identified a set of 37 risk components for the seven risk factors. For each component, we developed a taxonomy of terms that included boundary conditions, threshold values, risk indicators, and mitigation strategies. This structure allowed us to consolidate the risks associated with NASA-level decisions and develop a framework for considering potential mitigation strategies available to decisionmakers.

Developing such a methodology with disparate risk factors requires normalizing the factors so they can be compared through a common analytical lens. In a numerical sense, it requires having common units in the terms that are compared such that one can understand the comparative risk burden between, for example, political risks and technical risks or organizational risks and supply-chain risks.

In such an analysis, absolute values are far less important than relative comparisons of those values.[1] Such normalization allows one to assess where among the seven risk areas NASA leadership should have the most concerns and ultimately where

---

[1]   In the classical risk assessment format, when risks are represented by the likelihood of some outcome occurring (e.g., 10 percent of five fatalities), absolute values are very important. Once these risk values have been normalized to a common ordinal scale (e.g., 1-to-5 scale), the values become less descriptive and therefore lack the actionable aspect of their classical counterpart.

resources could or should be allocated to mitigate risks and improve chances of mission success.

The centerpiece of the normalization process is the analysis of the risk factors and their respective components. Each component must be analyzed to identify the metrics associated with the risk factor, the evaluation methods that can be employed to analyze them, and the mitigation measures that can be undertaken to alleviate the risks.

Components must include the boundary conditions that define the acceptable limits of each component. Ranges with maximum and minimum limits must be established to serve as these boundary conditions, where relevant.[2] Values or assessments that exceed the boundary conditions are, by definition, unacceptable, and therefore cause for either rejecting the entire program under consideration or incorporating mitigation measures.[3] Each component must be analyzed to ensure that all boundary conditions are eventually met. Should a component boundary condition not be met, the risk should be deemed unacceptable for the entire risk factor.

Evaluation methods, and the uncertainties attached to them, should be identified for each component. Undoubtedly, methods and uncertainties will vary among risk factors and within the individual components. Some will lend themselves to quantitative measures, while others will be analyzed qualitatively. As such, a mix of objective and subjective measures will likely be employed to evaluate each component; therefore, one should expect that a mix of empirical data, models, expert judgment, and historical data will form an important part of the analysis input.

The definitions for each risk and its boundary conditions, thresholds for normalization, and mitigation strategies will vary for each case under consideration. For example, in the cases we considered—cancellation of the Space Shuttle program and commercialization of transport to the International Space Station (ISS) as a previous decision and the Cislunar Habitat as a future decision—the methodology remained constant but the data elements within the methodology all changed to reflect the differences in these two case studies.

Findings in Brief

This report describes a methodology for assessing NASA-level risk. Our study incorporated a variety of factors to be considered in such an assessment. It further demonstrated that critical decisions could be identified and analyzed using this methodology, and that the methodology could be incorporated into analysis at various levels of fidel-

---

[2]   For example, if the value of the cost growth component is less than 5 percent, this may be mapped to the risk scale value of 0 (low), while 5–10 percent may be mapped to a 1 (some), all the way to where 100 percent would be mapped to a 5 (unacceptable).

[3]   Mitigation measures are operational strategies that may be taken to reduce the risk.

ity—including going into deeper layers of the problem when more-specific data are available. The methodology can also support sensitivity analysis.

We developed a set of factors for this study that define organizational-level risks associated with major decisions. Each factor represents an area where NASA faces risk. This provides a structured way to consider NASA-level risk, regardless of the issue under consideration. While these elements—the risk factors, their components, and their associated indicators and metrics—are not independent, they represent a fairly complete articulation of the elements and dependencies to be considered in a complex organizational risk assessment:

- **supply-chain risks**, including stability of sources and alternatives (domestic and foreign) for components, materials, and services; cost and schedule risks associated with maintaining supply chains or switching to alternative sources; quantity flow; and quality management
- **cost and schedule risks**, including at the program/project level and at an institutional level across multiple programs or directorates, as well as risks from budget instabilities (both in the short term, from continuing resolutions, and in the long term, from unstable long-range budgets)
- **human capital risks**, including to civil-servant and contractor support services, demographic issues, training, skill mix, quality, and flexibility and adaptability to changing missions
- **organizational and managerial risks**, including NASA-specific considerations such as the cultural and managerial tensions between headquarters and research centers, as well as risks to future missions requiring substantive changes (tolerance of risk, outsourcing, partnerships, external reliance)
- **external dependency risks**, including criticality to programs or missions, partnership and funding approaches for long-term stability, and insource/outsource trade-offs in cost and flexibility
- **political risks**, including reliance on foreign entities (e.g., Russian rocket engines), congressional restrictions for consolidation or alternative acquisition approaches, and strategy instability and priority shifts between presidential administrations
- **technical risks**, including alternative ways to adjust portfolio content to ensure robustness under uncertainty, cross-project reliance effects, risk mitigation in roadmaps, and risk-reward strategies (e.g., high-risk, high-payoff versus low-risk, low-payoff), and uncertainties from forecasting technology development.

As such, this methodology provides a comprehensive approach for guiding staff and decisionmakers through a structured and repeatable process for assessing risk. It allows for deliberating various risks in a decision and reaching consensus about their likelihood, consequences, and dependencies. Through this process, NASA can examine a wide range of complex, multidisciplinary issues facing the agency.

Traditional risk assessment methods also can be employed for individual risk factors, indicators, and mitigation strategies. They will form the basis for analyzing these individual elements and serve as inputs to the methodology that allows for normalizing and assessing the NASA-level risk in a particular decision or set of decisions.

We sought to validate our methodology through case studies of a past and future NASA decision. While we demonstrated that it was feasible to conduct an analysis integrating multiple factors that contribute to risk and to compare multiple forms of risk, we did so using data that we developed and not expert elicitation, as we recommend in our full methodology. As a result, while we could validate our methodology, we could not validate the previous decision regarding cancellation of the Space Shuttle program and commercialization of transport to the ISS; nor could we identify with authority the risks associated with the future Cislunar Habitat decision.

Thoughts on the Methodology

The analysis firmly established that the proposed methodology is robust and highly adaptable for a variety of cases, questions, and issues that NASA might face in the future.

Overall Conclusions

Our research indicates that the benefits of using a structured methodology far exceed the final calculations that result. Stated more directly, the process is far more important than that result.

The insights gained from employing the methodology and having to examine boundary conditions and thresholds generated important understanding of the sensitivity of risks for the issue under consideration. Requiring consideration of not only risks but also mitigations at an early stage in the analysis can prompt a structured thinking process that allows for clearer and more-thoughtful decisions.

While decisions may not be changed—from Option 1 to 2, or from "yes" to "no"—as a result of employing this methodology, gaining better understanding of key sensitivities will undoubtedly be an important outcome.

The use of a normalization process allows comparison of disparate risk issues that contribute to important decisions. Without such normalization, decisionmakers would be presented with a number of risk factors using different scales (and likely different grading and weighting) with little ability to understand comparisons within or among them or the overall risk burden associated with a decision.

The normalization process allows for determining relative component risks, and even risk indicators and mitigation strategies. However, as noted, the outcomes of the normalization process should only be considered in a relative sense and not used for absolute comparisons. In other words, one may use our methodology to determine that Option A is better than Option B, but not that Option A is 5 percent better than Option B.

The framework described in this report could be used by NASA to respond to current and future requirements for enterprise risk management, such as Office of Management and Budget Circular A-123 and its requirements for enterprise risk management (Clark, 2016).

Robustness of the Risk-Informed Decision Methodology

The proposed methodology allows for tailoring to a wide variety of risk factors. The analysis also developed a taxonomy for thinking about NASA organizational-level risk. For this taxonomy, a wide variety of risk factors, risk factor components, risk indicators, and mitigation strategies can be considered. This allows for assessments of either unmitigated or mitigated risk. One can also attach costs (to include time, personnel, and dollars) to the mitigation strategies to develop a cost-informed set of approaches.

The proposed methodology allows for comparing options. The Cislunar Habitat case study, for example, shows how risks vary by three options (international development, public-private development, or NASA development). This, in turn, allows comparison of the overall risk for each option, and the particular risks that most differentiate the options. Using these comparisons, the options themselves can be considered and compared, providing decisionmakers with a structured way to consider the risks of each.

Requiring that the analysis of each risk factor and its components, indicators, and mitigations begin with an identification of boundaries provides a clear articulation of the limits, acceptable and unacceptable, involved in the analysis. This development process allows senior leaders to articulate boundary conditions upon which a risk is no longer acceptable, providing guidance to the analysis.

This structured approach to conducting risk assessments builds understanding of the interdependencies within an issue. While eliminating all interdependencies involved in an issue might be desirable from an analytical standpoint, doing so is unrealistic for the types of complex, multidisciplinary issues that NASA faces on a regular basis. Some dependencies may be eliminated by the choice of indicators for the risk factors and their threshold values for normalization. So, while one cannot eliminate these interdependencies, it is certainly possible to define, understand, and assess how various relationships among the risk factors, components, indicators, and mitigations interact. Such reflection will provide insights into how best to manage and mitigate any of the negative effects of interdependencies. Some risk factors and indicators may be dropped from consideration, but only after careful deliberation.

Each Risk Assessment Case Requires Tailoring to the Problem in Question

Each discrete risk assessment requires that the methodology be changed to reflect the particulars of the issue under consideration. Each of our case studies, for example, used their own threshold values, weights, and scores for each of the risk indicators and mitigation strategies. These do remain constant, however, for considering options within the same case, such as the three options we consider in the Cislunar Habitat case study.

Improving the Quality of the Analysis Requires Resources

The quality of the risk assessment is directly related to the quality of the inputs for the assessment. Access to actual data, subordinate risk calculations (e.g., technical assessments of components within programs), and expert elicitation would improve the risk assessments. Employing more-rigorous methods—such as probabilistic risk assessments, earned value management, and decision trees—would improve assessment of risk values.

Significant resources may be required to develop the inputs for this methodology, but once those inputs are in hand, the only significant remaining task is eliciting weights from decisionmakers. The rest of the analysis could be automated using prepared Excel worksheets. Inputs for each of the risk factors may be obtained from different experts in each area. Therefore, while resource-intensive, much of the work may occur concurrently.

Separating the Building of the Methodology from the Decisionmaking

Our methodology calls for inputs from both subject-matter experts in assessing levels of risk and decisionmakers in determining the levels of acceptability for risk. To prevent bias from affecting the methodology, input from subject-matter experts and decisionmakers should be collected separately. Both are essential to the process, but decisionmakers must be able to synthesize the material and think at a high level of abstraction about the results, with only minimal input from subject-matter expert assessment. Providing this would create an independent role for the decisionmaker.

Important Caveat on Methodology Validation

Our case studies validating the methodology employed internal subject-matter experts to introduce realism into the risk assessment data that we used, but we did not have access to actual data associated with these decisions. As such, our case studies should not be considered authoritative from a substantive standpoint, only from a methodological perspective. For this reason, all the tables for our case study are marked as having notional data.

Moving Forward

The steps developed in the methodology provide a structured way to consider a risk-informed decision. While we were able to conduct abbreviated case studies validating it, conducting a more robust analysis for a future NASA decision is important. Given the work we have begun on it, applying this model to a more expanded analysis of the pending Cislunar Habitat decision may be most appropriate. Such work would provide further validation of the methodology, as well as additional insights into the risks (and options) associated with this decision.

Acknowledgments

This work benefited from the input and assistance of many people. Our sponsor, Patrick Besha from NASA's Office of Strategy and Plans, made this project possible. He provided opportunities for regular interaction with the NASA team, which greatly aided in the final analytical product. The cooperation of Todd Harrison and Nahmyo Thomas of the Center for Strategic and International Studies is greatly appreciated, as is that of Tom Cremins, associate administrator for Strategy and Plans, who facilitated interaction with some subject-matter experts.

We also appreciate the feedback from the other senior staff at NASA during our briefings. Their comments helped shape our study efforts and the final report.

We thank our administrative assistants, Alex Chinh, Jamie Greenberg, Michelle Horner, and John Tuten; and our management, Anita Chandra and Tom LaTourrette, for their contributions to this effort. We also are very appreciative of the comments from our reviewers, John Casani, Henry Willis, and one anonymous reviewer.

Abbreviations

AC	actual costs
CLSS	Cislunar Space Station
CPI	cost performance index
CV	cost variance
DoD	U.S. Department of Defense
EV	earned value
EVM	Earned Value Management
GAO	Government Accountability Office
GSFC	Goddard Space Flight Center
HQ	headquarters
IRL	Integration Readiness Levels
ISS	International Space Station
JPL	Jet Propulsion Laboratory
JSC	Johnson Space Center
JWST	James Webb Space Telescope
KSC	Kennedy Space Center
MRL	Manufacturing Readiness Levels
MSFC	Marshall Space Flight Center
NASA	National Aeronautics and Space Administration
NAVSEA	Naval Sea Systems Command
NPV	net present value
PV	planned value
SPI	schedule performance index
SRA	Society for Risk Analysis
SV	schedule variance

TC	technical capability
TCHA	Technical Capability Health Assessment
TRL	Technology Readiness Level
WFC	Warfare Center

CHAPTER ONE

Introduction

Overview and Study Terms of Reference

This study was conducted in response to a request from the National Aeronautics and Space Administration (NASA) to assist in gaining an understanding of how various disparate factors contribute to overall risk to project- or mission-level success that is associated with NASA-level decisions. The specific mandate was to provide NASA's Office of Strategy and Plans with a risk-informed decision support methodology that integrates risk factors and risk management approaches tailored to NASA's management, operations, and acquisition structures.

This report is written for a non-expert audience so that any practitioner or decisionmaker with any level of training could make use of the methodology. It therefore does not assume knowledge about risk or decision analysis on the part of the reader.

NASA confronts a variety of organizational-level risks within its programs. Comparing, contrasting, and mitigating these risks require developing a common lens through which to view them. Such an evaluation can provide opportunity to gain an overall understanding of the risks associated with NASA-level decisions. This study develops and describes such a methodology. Specifically, it resulted in the development of a single "risk framework" (hereafter called "risk assessment methodology") and includes the identification of various techniques for conducting risk assessments and the definition of seven risk factors identified as being important for NASA-level decision consideration.

RAND conducted a multidisciplinary effort that entailed a review of existing literature on risk assessment and management, identification of and definitions for overarching risk factors, development of a risk-informed decision methodology, and validation of this methodology using two case studies.

In conducting this study, RAND focused on the risk factors, best practices, and associated risk management approaches relevant to current NASA management, operation, and acquisition.[1] The intent was to inform NASA's ultimate interest in evaluat-

---

[1] Such risk factors, best practices, and associated risk management approaches may change based on NASA's decisionmakers and the space market at the time. The methodology allows for this type of tailoring.

ing specific risk postures of the agency; such analyses are beyond the scope of this study but could be follow-on projects for RAND.

This study leveraged RAND expertise across the applicable risk domains and consulted with subject-matter experts to search for other relevant approaches. RAND analyzed those approaches and integrated them into a multitiered, cross-functional methodology to facilitate review for their use in evaluating risk posture across various dimensions and levels of the NASA enterprise.

A methodology such as this is important given the breadth of the risk factors under consideration. This methodology, together with a brief overview of each of the risk factors and management approaches analyzed in the study, will be summarized in this report, along with extensive citations on each of these topics.

Any organization faces a range of risks to meeting its objectives. Risks cannot be eliminated, but they can be consciously managed and traded through various techniques that prioritize investments and policies. This study will examine the range of risk factors and existing approaches for managing those risks.

The risk factors RAND considered in the study included supply chains; acquisition costs and schedules; human capital or workforce issues (both civil servants and support contractors); organizational and managerial considerations (e.g., the issues arising from NASA's distributed management between mission directorates and research centers); dependency on external sources; domestic and international political risks (e.g., partnership instabilities, congressional equities) and technical risks, including information security. Of note, these were the original risk factors RAND identified in the proposal and that were agreed upon by NASA to be employed in the analysis. The seven risk factors utilized in the study effort are

- **supply-chain risks**, including stability of sources and alternatives (domestic and foreign) for components, materials, and services; cost and schedule risks associated with maintaining supply chains or switching to alternative sources; quantity flow; and quality management
- **cost and schedule risks**, including at the program/project level and at an institutional level across multiple programs or directorates, as well as risks from budget instabilities (both in the short term, from continuing resolutions, and in the long term, from unstable long-range budgets)
- **human capital risks**, including to civil-servant and contractor support services, demographic issues, training, skill mix, quality, and flexibility and adaptability to changing missions
- **organizational and managerial risks**, including NASA-specific considerations such as the cultural and managerial tensions between headquarters (HQ) and research centers, as well as risks to future missions requiring substantive changes (tolerance of risk, outsourcing, partnerships, external reliance)

- **external dependency risks**, including criticality to programs or missions, partnership and funding approaches for long-term stability, and insource/outsource trade-offs in cost and flexibility
- **political risks**, including reliance on foreign entities (e.g., Russian rocket engines), congressional restrictions for consolidation or alternative acquisition approaches, and strategy instability and priority shifts between presidential administrations
- **technical risks**, including alternative ways to adjust portfolio content to ensure robustness under uncertainty, cross-project reliance effects, risk mitigation in roadmaps, and risk-reward strategies (e.g., high-risk, high-payoff versus low-risk, low-payoff), and uncertainties from forecasting technology development.

Study Approach

Development of the study approach ensured that the team would begin from a position of a clear understanding of the state of the field of risk assessment to allow for developing a risk-informed decision analysis that captured the best of current thinking while pushing the envelope with respect to complex organizational risk designs. It also served to ensure that the methodology would be highly tailored to the NASA mission set. Figure 1.1 depicts the study approach.

In addition to building on its substantial experience in risk assessment, the team conducted a literature review, which allowed the study team to compile the most-relevant risk and decision assessment methods. It identified a number of methods for conducting risk-informed decision assessments, but most require determination of the probabilities associated with the consequences of concern. In many cases, these probabilities are unknown—and, sometimes, unknowable (i.e., not based on observable data). Organizations may be able to elicit subjective probabilities from subject-matter experts when these experts are readily available; in cases where they are not, it is common for organizations to make use of *risk indicators*, or knowable metrics that describe different aspects of the current state of the world that may provide an *indication* of likelihood that a certain consequence of concern to that organization will occur.[2] In essence, these risk indicators act as proxies for the actual risk associated with some event or set of circumstances. In this study, we make use of risk indicators to characterize high-level NASA risk.[3]

---

[2] As discussed further in Chapter Two, our use of the term *indicator* includes metrics related to project performance (e.g., cost and schedule growth metrics), as these may signify that there are underlying risks with the project.

[3] Our use of indicators in this study should not be confused with those used to develop prior distributions in Bayesian analyses. The methodology applies a frequentist approach to the decision problem. Mentions of updating previous results throughout this report, therefore, do not refer to Bayesian updating, but simply the act of the analyst reassessing the indicator based on updated information.

Figure 1.1
Study Approach

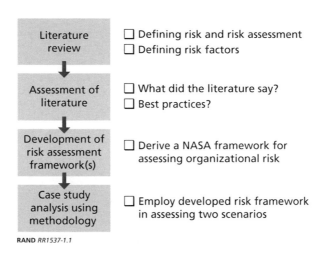

RAND *RR1537-1.1*

Thus, the literature review focused on defining risk-informed decision variables at three levels: the overall risk *factors*, each factor's *components*, and each component's risk *indicators*. Risk factors and components were identified by RAND and agreed upon by NASA to serve as the elements for an organizational risk-informed decision support methodology. We develop seven risk factors, each with multiple components—and within those, a set of indicators that suggest the level of risk (i.e., the likelihood that a certain consequence of concern will occur) associated with that area. We chose this method of risk assessment over a number of others described in Chapter Two (e.g., risk matrices, probability distributions) because many probabilities were unknown and it was not feasible in the current effort to find appropriate people and conduct structured expert elicitation exercises to estimate probabilities for the seven risk factors. The information compiled in this literature review phase served as a foundation for the remainder of the study.

In the next phase, the information was analyzed and best practices and methods for risk evaluation were identified. These elements would form the basis for the development of the risk-informed decision support methodology.

The methodology began with the identification of 37 risk factor components. For each component, a taxonomy of terms was developed that included boundary conditions, threshold values, risk indicators, and mitigation strategies for considering each of the seven risk factors. This structure allowed for consolidating the risks associated with NASA-level decisions, as well as developing a methodology for considering potential mitigation strategies available to decisionmakers.

To allow for developing relative comparisons among risk factor components, a normalization process was established such that they could be compared through a common analytical lens. In a numerical sense, it requires having common units in the terms that are compared such that one can understand the comparative risk burden between, for example, political risks and technical risks, or organizational risks and supply-chain risks.

For each risk factor component, boundary conditions were established, risk indicators and mitigation strategies were developed, and threshold values were chosen upon which to map the values for the components to a risk scale from 0 (low) to 5 (unacceptable). For example, indicators for supply-chain management may include percentage of cost growth or schedule growth. Each component is then assigned a weighting between 0 (not important at all) to 5 (extremely important). Finally, the sum product of the risk scales and weights are calculated to obtain a *normalized value* for the specific risk factor. Once a normalized value has been calculated for every risk factor, the risk factors are then graphically displayed on a radar chart, allowing graphical comparisons to be made.

In the case study analysis section, two decisions—one past and one future—were considered. This served to test and partially validate the methodology. The previous decision considered was the cancellation of the Space Shuttle program and commercialization of the transport to the International Space Station (ISS), while the future decision considered was the Cislunar Habitat question. A full validation was not possible given the scope and time constraints of the study efforts. Rather, the goal was to test the limits of the methodology on real-world issues and begin to understand the robustness of the methodology.

Organization of This Report

The remainder of this report is structured as follows. Chapter Two provides definitions of risk assessment and a survey of the types of tools and methods that can be incorporated into risk-informed decision assessments. Chapter Three provides definitions for the seven risk factors, including the components, metrics, and mitigation factors for each risk factor. Chapter Four utilizes the literature search information and data compiled in Chapters Two and Three to arrive at developing a risk-informed decision analysis for examining NASA's organizational risk for specific decisions. Chapter Five presents the findings from the use of the methodology for two NASA decisions, one previous decision (Space Shuttle/ISS) and one future decision (Cislunar Habitat), which serve as validation of the methodology. Chapter Six provides the findings and conclusions, as well as next steps and potential areas for future analysis. Appendixes A through D provide further detail on cases for consideration, the two case studies we selected, and methodology. Finally, supplemental notional data for Appendixes B and C, available online, provide additional depth on the case studies and examples of the structured framework used in this report.

CHAPTER TWO

Risk Assessment and Risk-Informed Decisionmaking Methods

This chapter provides an overview of the most-relevant (for this study) and widely used methods for risk assessment and risk-informed decisionmaking. As such, it will provide a basis for the development of a NASA-level risk-informed decision analysis that allows for comparing, understanding, and mitigating risks likely to result from organizational-level decisions. The chapter includes a section on risk assessment methods that provides a few example approaches that may be used to obtain the lower-level inputs required for developing the methodology.

This chapter provides an overview of the approaches that ultimately informed the high-level decision methodology described in Chapter Four.

Risk Definition

The term *risk* may refer to a number of different concepts, depending on the domain and application. Thus, there is no one definition of risk. The Society for Risk Analysis (SRA) provides both qualitative and quantitative definitions of risk in their glossary (SRA, 2015). Qualitatively, risk is the "possibility of an unfortunate occurrence," or the "uncertainty about and severity of the consequences of an activity with respect to something that humans value" (SRA, 2015, p. 3). Quantitatively, risk is the "combination of probability and magnitude/severity of consequences," or the "combination of the probability of a hazard occurring and a vulnerability metric given the occurrence of the hazard" (SRA, 2015, p. 3). Similar definitions exist from a number of foundational sources, including the International Organization for Standardization (2009) and the International Risk Governance Council (2008), as well as many others.

Operationalizing these definitions into functional equations provides us with a basic formulation for simple cases:[1]

---

[1]   See this chapter's section on probability distributions for the more generalized formulation. The current form provides only for "expected risk."

$$Risk = Probability \times Consequence \qquad\qquad \text{Eq. 1}$$

where the *consequence* is some undesirable outcome and the *probability* is the likelihood of that outcome. A number of variations exist on such a formulation, as already mentioned, based on the domain and application. For example, the U.S. Department of Homeland Security (2011) uses the following variation:

$$Risk = Threat \times Vulnerability \times Consequence \qquad\qquad \text{Eq. 2}$$

where the *consequence* again is some undesirable outcome, the *threat* is the probability of an event occurring, and the *vulnerability* is the conditional probability that, if the event occurs, it will result in the consequence. In this way, the product of the threat and vulnerability variables in Eq. 2 is equivalent to the probability variable in Eq. 1. In both equations, risk is therefore generally provided in units of the consequence measure.

In any uncertain situation, there is also the possibility of more than one consequence occurring. For instance, a schedule risk might be a 50-percent likelihood of no delay *and* a 50-percent likelihood of a three-month delay. In this example, the mean schedule risk would be a 1.5-month delay.

Finally, some formulations of risk require a third variable (in addition to probability and consequence): that of (risk) exposure. Exposure is simply, "being subject to a risk source/agent" (SRA, 2015, p. 7). Risks in which the consequence variable is given as a rate (e.g., accidents per year, dose per kilogram) require information on the contact and duration of that risk. For example, consider a component that fails (the consequence) an average of twice per year (the probability). For an exposure period of four years, the component would be expected to fail eight times (the expected risk).

Risk Assessment Methods

In this section, we focus on examples of different methods for assessing a risk that may be used to obtain the lower-level inputs needed for the methodology, including use of risk matrices, probability distributions, and indicators. In addition, since many observational data may not be available for many of the risks (e.g., political risks) included in this organizational-level methodology, we briefly discuss the use of elicitation of risk values.

Risk Matrix

The use of a risk matrix allows for the characterization of a risk using two metrics: probability/likelihood and consequence/severity. Figure 2.1 shows one example of a simple risk matrix, where a risk is characterized for its severity on a scale from 1 (least severe) to 5 (most severe), and for its probability on a scale from 1 (least probable) to 5 (most probable). In some instances, these severity and probability levels are defined using qualitative descriptions (e.g., very likely, not at all likely), or more specifically using quantitative thresholds (e.g., between 90 and 100 percent, between 0 and 1 percent). Once a probability and severity are chosen, the risk is relegated to the associated box of the risk matrix. Every box in the risk matrix is assigned a risk level (e.g., high, medium, low) sometimes denoted using colors (e.g., red, yellow, green), as shown in Figure 2.1. The higher the risk level assigned, the more urgent and important that it be addressed.

Risk matrix approaches are widely used in defense and government settings (e.g., Garvey, 1998; U.S. Department of Defense [DoD], 2006; Federal Aviation Administration, 2009). Their popularity may be a result of how easy they are to comprehend, and how easily risk matrix values can be elicited when needed. Given their wide use, a major advantage of such a method is the universal familiarity. However, risk matrix approaches include a number of serious disadvantages. First, the probability and severity categories are vaguely defined in many instances, leading to potential inconsistencies in interpretation and placement of risks (e.g., Budescu and Wallsten, 1985; Cox, 2008). Even when quantitatively defined (e.g., between 90- and 100-percent likelihood), the choice of thresholds for each category (e.g., whether the category is 90–100 percent or 85–100 percent) may be chosen arbitrarily. The choice of the thresholds—as well as the choice of which boxes are high, medium, or low risk—is one that should be informed by engagement with stakeholders and decisionmakers and be related to their risk acceptance. Depending on the thresholds being chosen, risk matrices may have poor resolution, where two very different risks can be assigned identical ratings (Cox, 2008). A further disadvantage with the risk matrix approach is that these matrices are typically structured with only one consequence and one probability. In reality, most risks have multiple consequences, each with an assigned probability (e.g., a car accident can result in a fatality, an injury, damage to the car, or no damages or

Figure 2.1
Example Risk Matrix

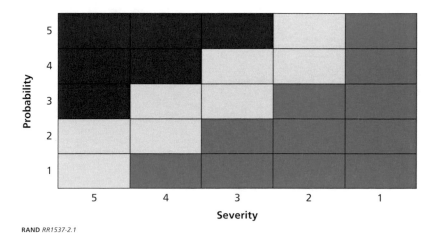

injuries). To properly represent the risk, all potential (classes of) consequences should be characterized. If only one consequence is characterized, there are further complications when different users of the matrix choose different consequences; one user may choose the "worst credible" consequence, another may choose the "worst possible," while still another may choose the "most likely" consequence. The variation may lead to a lack of consistency in the risk characterized. Risks with great uncertainty may also be misrepresented in a risk matrix. For instance, choosing to represent the "most likely" consequence in the matrix will not provide for the possibility that worse consequences could occur. In turn, the risk matrix makes it difficult to compare very uncertain risks, with a wide range of possible consequences, to less uncertain risks (Cox, 2008). A final limitation of the risk matrix is its inability to allow for aggregation of multiple risks, such as what would be needed in the case of system-level risk assessment. Does the aggregation of two medium-level risks equal a high-level risk? There is no consistent method to answer such a question.[2]

Probability Distributions

The use of probability distributions to characterize risks requires delineating all of the possible consequences of a risk and then assigning a probability to each consequence, such that those probabilities sum to 1. To properly build a probability distribution, all of the consequence values need to be measured in the same units (e.g., dollars, fatalities). Figure 2.2 shows an example of a probability distribution for the possible development costs of a new technology. The bar graphs represent the probability of each consequence, while the line graph represents the cumulative probability of the cost being less than or equal to that development cost.

In comparison to the risk matrix approach, the use of probability distributions to characterize a risk provides a fuller, more accurate picture of the risk (e.g., explicitly representing the tails of the distribution and its extreme values). Unlike the risk matrix, every potential consequence is represented; aggregation of risks is possible using rules for probability and/or Monte Carlo simulations (if the risks are considered to be independent[3]); probability and severity values must be defined, allowing for more-explicit interpretation of the risk characterization; and, in the case of continuous distributions, probability distributions remove the need for assigning (what may be arbitrary) probability and consequence categories. Overall, compared with a risk matrix approach, risks may be more accurately defined using probability distributions. On the other hand, probability distributions are more complicated to understand and to explain—

---

[2] Methodologies do exist, however, that allow analysts to tailor the aggregation formula to the specific context and to show the consequences of accepting different aggregation schemes. Some of the published examples are in evaluation of alternative military systems to be acquired (Davis, Shaver, and Beck, 2008; Davis and Dreyer, 2009).

[3] When not independent, an analyst must model the dependencies between risks to produce aggregated distributions.

Figure 2.2
Example Probability Distribution

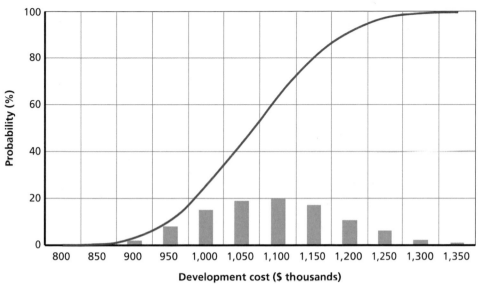

RAND RR1537-2.2

and they can be more burdensome to elicit values from individuals, if needed. Furthermore, building a distribution or aggregating risks may require quantitative analysis.

One of the most important advantages of using a probability distribution to characterize risk is its flexibility in presentation—and, in turn, interpretation. A distribution such as that shown in Figure 2.2 may be presented as is, or simple calculations may be performed to compute an expected value for the risk; the 25th-, 50th- (median), and 75th-percentile values for the risk; standard error; and standard deviation of the risk. The preferred representation of the risk will depend on the decision being made, as well as decisionmakers' and stakeholders' preferences.

Indicators

When likelihoods related to risks are unknowable, individuals and organizations may develop and make use of a set of risk indicators that may act as a proxy for the overall risks of concern. These risk indicators provide a representation of the current state of the world, which provide indications of the likelihood that a certain adverse consequence may occur.

For example, organizations concerned with systems acquisition risk, such as NASA and DoD, have developed risk indicators that characterize the technological and manufacturing maturity of components being developed in an acquisition process. The two most widely used are the Technology Readiness Level (TRL) and Manufacturing Readiness Level (MRL) scales, (e.g., Morgan, 2008; Mankins, 2009; Office of the Secretary of Defense Manufacturing Technology Program in collaboration with the Joint Service/

Industry MRL Working Group, 2011; Mai, 2015). Similarly, the Integration Readiness Level (IRL) scale provides a measure on the maturity between systems (Sauser et al., 2009). Figure 2.3 shows the TRL scale and its acquisition phases. The concept behind this readiness level is that the current state of technology practice provides an indication of whether the technology will be successfully acquired and implemented.[4]

A number of project performance metrics may also be used as risk indictors, as they may signify that risks exist related to project performance. One set of metrics is developed as part of Earned Value Management (EVM), which compiles a number of indicators based on three project values:

- *planned value* (PV): the approved budget for accomplishing the work on the project within the schedule
- *earned value* (EV): the cumulative worth of the work completed up through a specific time, calculated as the amount budgeted for performing that work
- *actual costs* (AC): the actual costs incurred to accomplish the work on the project to a given point in time (Fleming and Koppelman, 2002).

Figure 2.3
Technology Readiness Levels

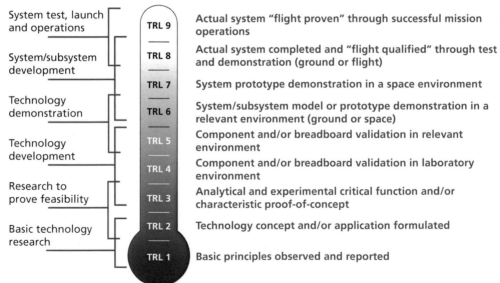

SOURCE: Office of the Secretary of Defense Manufacturing Technology Program in collaboration with The Joint Service/Industry MRL Working Group, 2011.
RAND RR1537-2.3

---

[4] While one could attempt to convert these readiness levels into probabilities of success or failure, they are commonly used "as is" in risk assessments. Given that they are "backward-looking" metrics, TRLs and MRLs are considered to be inferior, but more available, than such "forward-looking" metrics as those related to potential uncertainties.

A number of project performance metrics that can be used as risk indicators can be calculated based on these project values. These attributes include a number of variances:

- cost:
  - cost variance (CV): $CV = EV - AC$; measures the budgetary conformance of actual cost of the work completed
  - cost variance percent (CVP): $CVP = CV/EV$; measures the cost variance as a percentage of the earned value
  - cost performance index (CPI): $CPI = EV/AC$; efficiency ratio for cost variance
- schedule:
  - schedule variance (SV): $SV = EV - PV$; measures the conformance of actual progress to the schedule
  - schedule variance percent (SVP): $SVP = SV/PV$ or SV/EP; measures the schedule variance as a percentage of planned value/earned value
  - schedule performance index (SPI): $SPI = EV/PV$; efficiency ratio for schedule variance
- critical ratio (CR): $CR = CPI \times SPI$; indicator of overall project performance (may also be called the cost schedule index) (Anbari, 2003).

One other method for developing risk indicators, used specifically for business management, is the Balanced Scorecard approach. Kaplan and Norton (1996), who coined the term, argued that business management relied too heavily on financial accounting measures (i.e., only one type of attribute) and that a scorecard should be developed that takes other measures (i.e., attributes) into account. They proposed a scorecard with four components: finance, internal business, innovation and learning, and customers. Businesses would then develop a set of goals or objectives and place each objective into one or more of the scorecard components. Finally, metrics (i.e., attributes) would be developed that could be measured to evaluate how the business was doing on that objective within each of the four components. If an organization performs poorly on these metrics, it could signify that organizational risks exist. Therefore, these metrics may be used as indicators. Figure 2.4 shows an example of a balanced scorecard.

Elicitation of Risk Values

For many risk assessments, observed data that can inform the characterization of a risk (whether to build a probability distribution or determine the risk matrix values) are not available. In such situations, individuals and organizations may rely on their expert judgment to determine the values. This approach is widely used across defense, government, and industry organizations (e.g., Atwood et al., 2005; Morgan et al., 2006; Curtright, Morgan, and Keith, 2008; Ferdous et al., 2014; Bond et al., 2015; Markel et al., 2015) and, when appropriately applied, can lead to reliable and accurate risk estimates (Morgan and Henrion, 1990; Clemen and Winkler, 1999; Ayyub, 2001a;

Figure 2.4
Example of a Balanced Scorecard

Financial Perspective	
GOALS	MEASURES
Survive	Cash flow
Succeed	Quarterly sales growth and operating income by division
Prosper	Increased market share and return on equity

Customer Perspective	
GOALS	MEASURES
New products	Percentage of sales from new products
	Percentage of sales from proprietary products
Responsive supply	On-time delivery (defined by customer)
Preferred suppliers	Share of key accounts' purchases
	Ranking by key accounts
Customer partnerships	Number of cooperative engineering efforts

Internal Business Perspective	
GOALS	MEASURES
Technology capability	Manufacturing geometry versus competition
Manufacturing excellence	Cycle time, unit cost, yield
Design productivity	Silicon efficiency, engineering efficiency
New product introduction	Actual introduction schedule versus plan

Innovation and Learning Perspective	
GOALS	MEASURES
Technology leadership	Time to develop next generation
Manufacturing learning	Process time to maturity
Product focus	Percentage of products that equal 80 percent of sales
Time to market	New product introduction versus competition

SOURCE: Kaplan and Norton, 1996.
RAND RR1537-2.4

Ayyub, 2001b; Galway, 2007). During an elicitation, experts are prompted to make subjective judgments about uncertain quantities of interest. These experts use their best judgment, informed by historical experiences with the subject matter.

As an example, elicitations are commonly used to develop triangular probability distribution by eliciting the potential range of values a consequence can take on, as well as the most likely value. For instance, a schedule risk triangular distribution may be developed by asking experts how long it will take for a technology to be ready for deployment. An elicitation facilitator could ask experts for the least and most amount of time it could take, followed by the most likely time frame.

Experts who conduct risk elicitations (and those relying on their conclusions) should be careful of the heuristic processes (i.e., mental shortcuts) that may occur when they elicit subjective judgments (Kahneman, Slovic, and Tversky, 1982; Morgan and Henrion, 1990; Ayyub, 2001b; Galway, 2007; Hastie and Dawes, 2010). Availability, or the tendency to overestimate the probability of events that are easy to recall, and overconfidence, or the tendency to underestimate the uncertainty surrounding certain elicited quantities, are two heuristics that can commonly affect experts' risk elicitations (Morgan and Henrion, 1990; Hastie and Dawes, 2010). A number of established methods have been developed to ensure that the effects of heuristics during an expert elicitation are minimized, such as training experts on the common heuristic pitfalls during elicitations and following a systematic elicitation method (Morgan and Henrion, 1990; Ayyub, 2001a). A short review of expert risk elicitation methods and heuristic processes can be found in Appendix C of Bond et al. (2015).

Risk-Informed Decisionmaking

With risks assessed, decisionmakers may explicitly and implicitly use a number of decisionmaking methods and considerations to make decisions regarding whether the risk is acceptable as is or needs to be mitigated, the level of uncertainty they will tolerate surrounding the risk, and how risks should be ranked in order of concern. Here, we review these three concepts.

Acceptable Risk

Once risks have been characterized, the logical next step of any individual or organization being exposed to those risks is to determine whether they are at an acceptable level. Before embarking on any risk-reduction methods, individuals and organizations must decide whether risks are serious enough to undertake mitigations. These thresholds, or an individual's or organization's risk-acceptance level, may be established in a number of ways, including:

- The risk value falls below a set standard or defined probability.
- The costs of mitigations outweigh the risk.
- Scientific research finds the risk level to be minimally threatening to society.
- Public opinion warrants the risk.
- Policy indicates acceptability of the risk.

In each of these examples, it is implicit that the determination of what constitutes an acceptable risk level requires considering what benefits come along with undertaking the risk and the other options available. If no benefit exists, or if there are other ways to receive any potential benefits, the risk may be unacceptable at any level. In most cases, individuals and organizations weigh risks against benefits to determine thresholds. For example, the Environmental Protection Agency generally considers 10^{-6} to be an acceptable level of risk of death from lifetime environmental exposures (Kelly and Cardon, 1991). However, the Occupational Safety and Health Administration generally assumes that the acceptable level of occupational risk over a lifetime is

10^{-3} (e.g., Sadowitz and Graham, 1995; National Research Council, 2004). The stated reason for this difference is that occupational risks are typically taken on voluntarily and for compensation. Similarly, sufferers of terminal diseases generally find the risks of side effects from medical treatments acceptable if the benefits lead to a cure, yet the same side effects are not acceptable if treatments do not lead to a cure.

Risk Attitudes

The risk attitude of individuals is sometimes referred to as their risk tolerance, or an individual's preferences regarding uncertainty (e.g., Pratt, 1964; Arrow, 1965). The concept of risk tolerance is closely related to (expected) utility theory, which states that individuals prefer alternatives that provide them with greater utility (e.g., von Neumann and Morgenstern, 1944). A simple example may explain the concept of risk tolerance. An individual is given a choice of payment for a good or service: Pay $50 or accept the outcome of a coin toss with a 50-percent chance of paying $100 and a 50-percent chance of paying $0. While the expected payment in both scenarios is $50, a risk-averse individual (someone with a low tolerance for risk) who dislikes uncertainty may choose the sure payment, while a risk-seeking individual (someone with a high tolerance for risk) may choose the coin toss. A risk-neutral individual would be indifferent to the choice. Individuals with different risk attitudes will value a risk differently. For additional information on subjective expected utility theory and risk attitudes, see Clemen and Reilly (2001) and Hastie and Dawes (2010).

Risk Comparison

Each of the methods explored in this chapter provides a means for characterizing individual risks. Individuals and organizations often need to compare sets of risks to establish priorities for mitigation or risk monitoring. In this section, we describe how each of the methods presented above can be used to compare and prioritize risks.

Risk Matrix

Risks that are characterized using the risk matrix can be compared by placing each in its respective matrix box, as shown in Figure 2.5. The figure demonstrates that Risks A, B, and E are more serious than Risks C and D, which are more serious than Risk F. We also may be able to determine that Risk A is more serious than Risks B and E. However, the matrix does not discern between Risks B and E. People assess high-probability, low-severity events and low-probability, high-severity events very differently (e.g., Fischhoff et al., 1978), and the risk matrix shown in Figure 2.5 cannot account for that. That is, Risks B and E are different but have the same rating in the risk matrix (red); thus, decisionmakers may erroneously see these risks as being equivalent (Cox, 2008). Processes can be designed to distinguish between these two types, such as the Hazard Risk Index used by the U.S. Navy, in which each box within the matrix is given a ranked priority (Massingham, 2010). Another concern when comparing risks using a risk matrix is individual bias: Two risks that are assessed by different

Figure 2.5
Risk Comparison Using the Risk Matrix

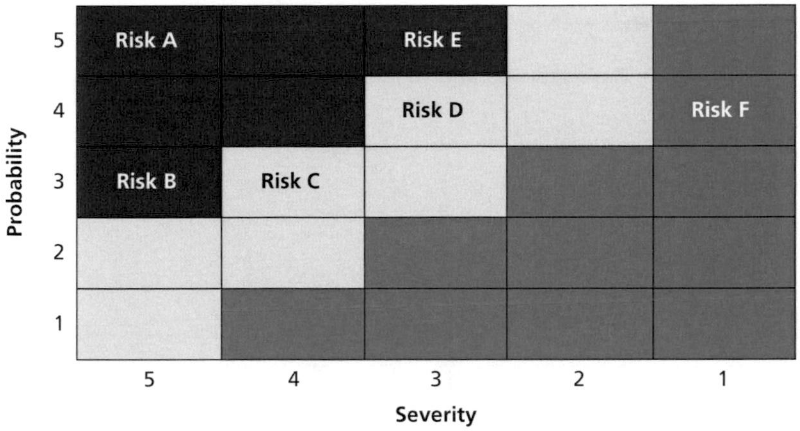

individuals and placed in the same box may or may not actually be equal. Because of the potential for inconsistencies in characterizing the risks that we have already discussed (e.g., vague category descriptions leading to different interpretations), risks assessed by two different individuals, or even the same individual at two different times, may not be comparable. Finally, the risk matrix cannot accommodate comparisons of disparate risks unless given a common consequence metric. For instance, a technical risk (e.g., component failure) cannot be directly compared with a schedule risk (e.g., delay of program delivery) unless both are converted to a similar metric, such as monetization. In this case, we could assess the monetary damage of a component failure and provide a cost to each increment of program delay. The consequence metric would become dollars and then both risks could be assessed in those terms.

Probability Distributions

Similar to the risk matrix, probability distributions cannot accommodate comparisons of disparate risks unless given a common consequence metric. But, unlike the risk matrix approach, probability distributions provide more information that can help an individual or organization to compare the seriousness of risks. For example, Figure 2.6 shows the probability distributions as bars representing the 50-percent confidence interval between the 25th and 75th percentiles for three cost risks A, B, and C. Risks A and C clearly are less serious than B, which has the most negative lifetime net present value (NPV). When comparing risks A and C, we see that the mean NPV (denoted as an X in the figure) is slightly higher for A, but the median NPV (denoted as a red line) is better for C. The separation between the mean and median values for C denotes that the probability distribution is skewed to the left (most of its density is in the higher NPVs, but there are a few scenarios in which a very negative NPV could occur). This is also represented by the larger 50-percent confidence interval of Risk C when compared with A. All of this

Figure 2.6
Comparison of Probability Distributions

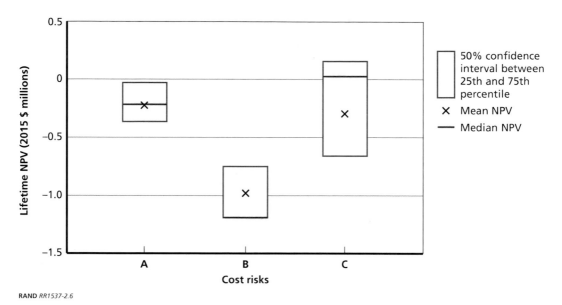

information can be displayed by a simple bar chart when using probability distributions, providing a decisionmaker with increased information with which to make a decision. In this case, the decisionmaker may want to avoid the potential for very negative NPV scenarios, such that they may see Risk C as being more serious even though it has a greater likelihood of also having better NPV values than Risk A.

Multi-Attribute Methods

While both the risk-matrix approach and use of probability distributions rely on a common consequence metric, a number of methods exist that can handle comparisons of disparate risks in which the consequence metrics are different (Keeney and Raiffa, 1976; Keeney, 2009). These multi-attribute methods require that the risks first be characterized by a set of attributes, or characteristics that are common among the risks. Attributes are chosen based on the preferences and values of decisionmakers and stakeholders and are relevant to the risk question. Every risk must be able to be defined for every attribute. The risks are then characterized, either quantitatively or qualitatively, according to these attributes. Qualitative characterizations must be defined with an ordinal set of values (e.g., high, medium, and low).

A decisionmaker may then compare these risks across the set of attributes to inform the prioritization decision. As an example, consider two hypothetical disparate risks a commercial aviation organization may be concerned with: a system safety risk (e.g., helicopter crash due to material failure) and an occupational health risk to the organization's employees (e.g., injury of workers on the manufacturing line). If the organization does not prefer to monetize all risks, or if its concerns include non–risk related or qualitative attributes, it may wish to compare the risks on a set of common attributes. An example characterization of the two risks is shown in Table 2.1.

To make these risk characterizations easier to compare, they may be normalized on a scale from 0 to 1, where a 0 represents the best value an attribute could take on, and a 1 represents the worst. One means for standardizing quantitative values is according to the following formula:

$$v_i = \frac{x_i}{\max_i - \min_i} \qquad \text{Eq. 3}$$

where, for the ith attribute, x_i represents the attribute value, max_i and min_i represent the maximum (worst) and minimum (best) values that attribute i can take on and v_i represents the standardized value. Eq. 3, a linear normalization approach, is best for attributes where the benefit associated with a marginal increase in the attribute is uniform across all possible values of the attribute (Keeney, 1980). Other equations (usually exponential or logarithmic) are common for other situations, such as where the benefit associated with a marginal increase in the attribute decreases as the attribute value increases (i.e., saturation) (Keeney, 1980). An exponential utility equation also may be applied to represent certain risk attitudes, such as risk aversion (Holloway, 1979; von Winterfeldt and Edwards, 1986).

Table 2.2 presents the normalized values, as well as hypothetical minimum and maximum values for the attribute upon which the normalized values are based. Note that we have assigned values to the qualitative uncertainty attribute such that low = 0, high = 1, and medium is determined to be halfway between those values at 0.5. For more information on determining attribute values for qualitative attributes, see Keeney (1980).

These risks may also be visualized to allow for better comparison. One visualization technique is the radar chart—also known as a polar chart, web chart, start plot, spider chart, or cobweb chart. It is a graphical decision support tool displaying data of more than one attribute in a two-dimensional chart. Each attribute is represented by radii that form a weblike design. Risks may then be plotted along the radii as spokes, according to their values; then, a line is drawn to connect the spokes for each risk. Figure 2.7 displays a radar chart for the two hypothetical risks.

Table 2.1
Example Characterization of Two Disparate Risks

Attribute	System Safety Risk	Occupational Health Risk
Maximum possible cost ($ millions)	10	2
Most likely cost ($ millions)	5	0.01
Expected frequency per year	0.25	10
Expected number of fatalities per occurrence	1	0.05
Uncertainty with risk estimate	Low	Medium

Table 2.2
Example Normalized Values of Two Disparate Risks

Attribute	Value Range: [Minimum, Maximum]	System Safety Risk	Occupational Health Risk
Maximum possible cost ($ millions)	[0, 10]	1	0.2
Most likely cost ($ millions)	[0, 4]	0.5	0.0025
Expected frequency per year	[0, 25]	0.01	0.4
Expected number of fatalities per occurrence	[0, 2]	0.5	0.025
Uncertainty with risk estimate	[Low, High]	0	0.5

From Table 2.2 and Figure 2.7, we see that the system safety risk is less serious in regard to the frequency and uncertainty attributes, but the health risk is less serious in terms of the two cost attributes and the expected number of fatalities per occurrence attribute. A decisionmaker who believes that each attribute affects risk prioritization equally would then likely see the system safety risk as more serious. However, decisionmakers and stakeholders may value some attributes more than others when making decisions about risk. Furthermore, certain levels of one or more attributes may be considered unacceptable and may lead to the entire risk being deemed unacceptable regardless of the evaluation of the other attributes. To represent these types of considerations, multi-attribute utility functions (Keeney and Raiffa, 1976; Keeney, 1980; von Winterfeldt and Edwards, 1986) are commonly used to provide a holistic assessment of the risk. These functions allow a decisionmaker to distill a multifaceted risk into one number—in this case, one that would represent the overall seriousness of the risk.

Figure 2.7
Radar Chart of the Two Hypothetical Risks

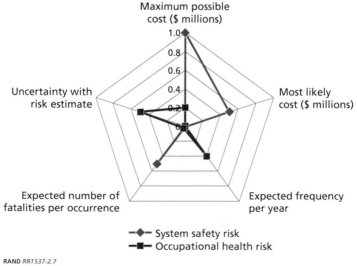

RAND *RR1537-2.7*

These multi-attribute utility functions can take on a number of different aggregations or formats. Most utility functions commonly used for this purpose include weights (that sum to 1) assigned to each attribute that represent the respective importance of that attribute to the decisionmaker. If we assume that the attributes are independent of one another, the aggregated utility is expressible as an additive model or multiplicative model, both linear combinations of weighted attribute utilities (Keeney and Raiffa, 1976). If attributes are assumed to be interdependent, the function would need to contain interaction coefficients for the attribute weights (Keeney, 1974). In addition to linear models, utility functions may take on nonlinear forms. An applicable nonlinear form includes that of threshold values for specific attributes to indicate their acceptability levels. For example, if the commercial aviation organization exemplified in in Table 2.1 decided that a single risk resulting in more than $8 million in damages was above the threshold of acceptability, the system safety risk with a maximum possible cost of $10 million would be deemed unacceptable regardless of the values for the other attributes.

A number of studies have shown that additive linear models can approximate other, more-complex modeling forms relatively well (Dawes and Corrigan, 1974; von Winterfeldt and Fischer, 1975; Dawes, 1979). The weights and normalized values of this model are used to calculate an overall risk score using the formula:

$$r = \sum_i v_i w_i \qquad \text{Eq. 4}$$

where, for the ith attribute, w_i is the weight, v_i is the normalized value, and r is the overall risk score. The weights assigned to the attribute may or may not have a drastic effect on the risk score. Table 2.3 shows two different weighting schemes that have been applied to the hypothetical risks from Table 2.2. In the equal weighting scheme, the commercial aviation organization denotes that all attributes are of equal importance. In the alternative weighting scheme, this organization suggests that it is most concerned with repeated events occurring, possibly because it wants to minimize its negative publicity. Table 2.4 shows the risk scores for both risks under the two weighting schemes. Under the equal weighting scheme, the organization would be most concerned about the system safety risk, while the alternative weighting scheme shows the health risk as being more serious.

Table 2.3
Weighting Schemes for the Hypothetical Risks

Attribute	Equal Weighting Scheme	Alternative Weighting Scheme
Maximum possible cost	0.2	0.1
Most likely cost	0.2	0.1
Expected frequency per year	0.2	0.4
Expected number of fatalities per occurrence	0.2	0.2
Uncertainty with risk estimate	0.2	0.2

Table 2.4
Risk Scores for the Hypothetical Risks

Scheme	System Safety Risk	Occupational Health Risk
Equal weighting	0.40	0.23
Alternative weighting	0.25	0.29

Weighting attributes indicates the magnitude of one attribute's importance relative to another. There are a number of methods that can be used to reliably elicit the weights needed for a multi-attribute utility approach (e.g., Edwards and Barron, 1994). The simplest method (the Ranked Sum method) involves eliciting a ranking for a set of n attributes from 1 (least important) to n (most important) and then assigning each attribute a weight equal to its rank divided by the sum of all the ranks (Barron and Barret, 1996). Another common method uses absolute pairwise judgments (Barron and Barret, 1996), where the weight of all attributes is developed relative to one specific attribute. For example, consider three weights w_1, w_2, and w_3. Using an absolute pairwise judgment method, we may assert that attribute 2 is twice as important as attribute 1, and attribute 3 is three times as important as attribute 1. The following set of equations can then be solved:

$$\left[\begin{matrix} 2w_1 = w_2 \\ 3w_1 = w_3 \\ w_1 + w_2 + w_3 = 1 \end{matrix} \right] \qquad \text{Eq. 5}$$

These weights would then become: $w_1 = 0.17$, $w_2 = 0.33$, $w_3 = 0.5$.

A number of other methods are available for calculating weights, including the swing weighting method, the ranked exponent method, and the ranked reciprocal method (Stillwell, Seaver, and Edwards, 1981; Barron and Barret, 1996). It is also worth noting that there is disagreement in the literature about the best or most accurate method for eliciting weights, as well as whether it is a good idea. Most studies show that equal weights (and even random weights) perform better than those chosen by experts (e.g., Dawes and Corrigan, 1974; Wainer, 1976). However, some studies do show the converse (e.g., Schoemaker and Waid, 1982). There is agreement that the most important factor in using the multi-attribute utility method is to get the attributes right (Dawes and Corrigan, 1974). A good compromise may be to use both equal and elicited weights. If the risk rankings are the same between them, decisionmakers can be more confident in the results. Similarly, if decisionmakers disagree on weighting schemes, both alternative weightings can be constructed, thereby highlighting whether it is important to choose between such perspectives.

CHAPTER THREE

Definition and Discussion of the Risk Factors and Components

As detailed in Chapter Two, a number of methods are available for conducting risk assessments. Most, however, require determination of the probabilities associated with the consequences of concern. In many cases, there are no data or methods to assess these probabilities. Organizations may be able to elicit probabilities from subject-matter experts if such experts are available. If they are not, organizations may use *risk components*, each of which has a set of problem-specific *risk indicators*. These risk indicators are knowable metrics that provide an *indication* of how likely a certain consequence of concern is to occur. In essence, these risk components act as the composition for the actual risk associated with some event or set of circumstances. In this chapter, we describe a set of risk components for seven risk factors:

- supply-chain risks
- cost and schedule risks
- human capital risks
- organizational and managerial risks
- external dependency risks
- political risks
- technical risks.

These risk components and their associated indicators (where applicable) have been tailored for the case studies presented in Chapter Five. Additional components or indicators may exist for each risk factor but may not be relevant for our case studies. Our methodology does allow the choice of other components and indicators where appropriate.

The risk areas listed above may have overlapping elements but each has a distinct theme. We explore each area as it relates to overall NASA risk while accounting for overlap among risk areas. Combined, the risk factors listed above form a portfolio of risk assessment.

Supply-Chain Risks

Supply-chain risk is the risk to NASA that services and materials cannot be obtained from a stable, high-quality source in a timely manner. The risk considers alternative suppliers and their stability and quality. While many organizations face supply-chain risks, the risks faced by NASA can be unique due to the unique nature of NASA's work.

NASA operates a wide range of equipment and facilities varying in age and condition. Finding a stable source of equipment is important for NASA. The history of the retired Space Shuttle program illustrates the supply-chain risks faced by NASA programs. The original shuttle design is rooted in the 1970s. The first shuttle, *Enterprise*, rolled off the production line in September 1976 (Slovinac and Deming, 2010). The final shuttle flight, that of *Atlantis* on STS-135, took place in July 2011. During the course of the program, the shuttle design had no drastic changes, meaning most of the design characteristics of *Enterprise* were present through the rest of the fleet for the entire duration of the program. Through the course of the program, the supply of parts was necessary to ensure missions could be accomplished. Some companies supplied parts through the duration of the program, while others were present and supplied parts for only part of the program's duration. This meant that alternative sources for components and equipment needed to be identified and the parts tested, certified, and used. In 2002, NASA began searching the Internet for sources of out-of-production computer parts for the shuttle (Broad, 2002). Relying on eBay for critical parts of the shuttle represented a supply-chain risk for NASA.

A more recent incident also illustrates the risks that supply chains pose. In June 2015, a SpaceX Falcon 9 rocket exploded just after launch. The cause was determined to be a strut inside the second stage of the rocket that failed due to material and manufacturing deficiencies (Stone, 2015). Here, a single, improperly manufactured component that was not detected in the quality-control process led to mission failure.

Supply-Chain Risk Components

Supply-chain risks can be characterized using a number of components and metrics. Some of these metrics and indicators can be quantified, while others may only be able to be characterized qualitatively. To ensure missions are able to operate as planned, a number of components should be analyzed as part of supply-chain risk. These include the following items:

- availability of materials
- availability of services
- stability of sources for components and equipment
- availability of alternative sources (domestic and foreign) for components and equipment
- quality management.

Availability of Materials

The first component is the availability of needed materials. This is assessed in two different ways. First, do the materials exist somewhere? Second, how easily can they be accessed?

Availability of Services

The availability of services can also be assessed in two ways. First, do the services exist somewhere? Second, how easily can they be acquired?

Stability of Sources for Components and Equipment

While materials may be available from a given vendor, the vendor may not be stable, or the demand for the materials may not be great, thereby threatening the stability of a production line. Are there sources available for products and services? Can the necessary amount of products be acquired? One should assess this risk by assessing the quantity of the material.

Availability of Alternative Sources (Domestic and Foreign) for Components and Equipment

Are there sources available for products and services even if they are not the original source of the products and services? Are the sources domestic or foreign? NASA should develop a risk assessment of both foreign and domestic suppliers.

Quality Management

Keeping high standards is important to NASA. Ensuring the quality of products is an important step in the supply chain, for the vendor, and for NASA. NASA should consider whether the products and services are up to standards for the given mission or use.

Cost and Schedule Risks

Cost and schedule increases pose the risk of missing a mission window or even program cancellation. The development of realistic costs and schedules is a necessary task at NASA. Cost and schedule can be traded within a set project scope; e.g., managers may choose to extend a project over time as a way to reduce its immediate cost. To perform these trade-offs, managers must use measurements at the mission, program, and institutional levels.

Of the risk factors discussed in this report, cost and schedule risk may have the most available literature. The cost and schedule risk discussed here are two of the three sides of the cost, schedule, and capabilities trade-off triangle. We assume that capability is fixed, and the only trades available to NASA are cost and schedule. That is, we explore the risks to the project in overrunning its schedule or its budget. We use the same approach for research and development programs and capital projects, though different components may exist for each.

NASA's Office of the Inspector General released a report in 2012 detailing some cost and schedule increases that have been observed in the agency:

> For example, in 1977 NASA estimated that it would complete development of Hubble in 1983 at a total cost of $200 million; however, the telescope was not completed until 2 years later at a cost of approximately $1.2 billion. More recently, MSL [Mars Science Laboratory] launched 2 years behind schedule with development costs that increased 83 percent, from $969 million to $1.77 billion. Similarly, in 2009 NASA estimated JWST [James Webb Space Telescope] would cost $2.6 billion to develop and launch in 2014; however, it is now projected to cost $6.2 billion to develop and launch in 2018. Cost increases and schedule delays on NASA's projects are long-standing issues for the Agency. A 2004 Congressional Budget Office study compared the initial and revised budgets of 72 NASA projects between 1977 and 2000. The initial budgets for these projects totaled $41.1 billion, while their revised budgets totaled $66.3 billion, a 61 percent increase. Moreover, since its first annual assessment of NASA projects in 2009, the Government Accountability Office (GAO) has consistently reported on cost growth and schedule delays in the Agency's major projects. For example, in its 2012 assessment GAO reported an average development cost growth of approximately 47 percent, or $315 million, much of which was attributable to JWST. As GAO noted, cost and schedule increases on large projects like JWST can have a cascading effect on NASA's entire portfolio. (Office of Audits, 2012, p. i)

The cost and schedule increases for the programs detailed in this NASA report are examples of this risk category. The cost and schedule delays faced by these NASA programs resulted in the workforce being used longer than planned and budgets being used for things that were not planned. These had cascading effects through the agency. The same report concluded that while optimism in exploration is positive, optimism in schedule and cost is detrimental to the agency. It also concludes that uncertain funding plays a part in the cost, schedule, and performance of NASA programs and projects.

> Project managers stated that they routinely struggle to execute projects in the face of unstable funding, both in terms of the total amount of funds dedicated to a project and the timing of when those funds are disbursed to the project. Both forms of funding instability can result in inefficient management practices that contribute to poor cost, schedule, and performance outcomes. (Office of Audits, 2012, p. iv)

An example of cost and schedule risks comes from the birth of the Space Shuttle program, and the delays faced by the program prior to the first flight of *Columbia* in 1981. Original plans called for the shuttle to begin flying in 1977, four years earlier than what was achieved (Armagh Planetarium, 2013). This would have allowed the shuttle to visit Skylab, the U.S. space station in orbit at the time, and to boost its orbit (Figure 3.1), preserving it as a viable laboratory for NASA.

Figure 3.1
Conceptual Space Shuttle *Columbia* Mission Boosts Skylab into
Higher Orbit

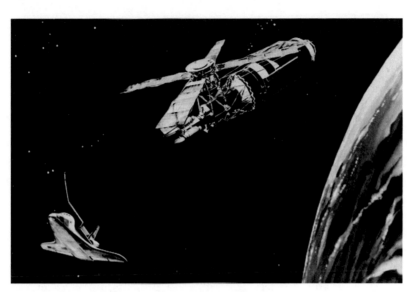

SOURCE: NASA, reprinted in Armagh Planetarium, 2013.
RAND *RR1537-3.1*

Instead, the cost increases and schedule delays resulted in the loss of Skylab; without a boost, Skylab's orbit decayed and it reentered Earth's atmosphere on July 11, 1979 (Hanes, 2012). Had the cost and schedule delays on the Space Shuttle program not occurred, it would have begun with an operational and habitable space station already in orbit. Skylab was a sunk cost that represented an opportunity for several more years of on-orbit research. Its loss left the United States without a habitable space station until November 2, 2000, when Expedition 1 docked with the ISS and began permanent human presence on orbit (NASA, 2015b).

Cost and Schedule Risk Components

Cost and schedule risks can be characterized using a number of components and metrics. Some of these metrics and indicators can be quantified, while others may only be assessed qualitatively. Components include:

- cost associated with maintaining access to needed resources
- schedule associated with maintaining access to needed resources
- budget stability
- program reliance on set cost and schedule
- insource/outsource trade-offs in cost and flexibility.

Cost Associated with Maintaining Access to Needed Resources

These are the costs needed to maintain access to needed resources, and deal more with the access aspect than with resource availability and redundancy of supply lines. Sometimes a supply chain can be supported with minimal investment. Sometimes it needs large investment.

Sometimes alternative suppliers need to be identified. There is a cost associated with switching to alternative sources—with forced changes often costing more than at-will changes.

Schedule Associated with Maintaining Access to Needed Resources

Active supply chains should be able to meet schedule demands, but sometimes these demands are unattainable. Schedule is related to cost in that extension of the schedule brings additional cost. There is also a schedule risk associated with switching to alternative sources—with risks differing by whether the change is forced or at-will.

Budget Stability

The stability and forecast of the budget directly relates to how far out NASA can schedule to spend money on work. Federal funds must be used within a relatively short period of time; thus, many NASA programs and projects are well-detailed for two years' worth of budget. Within those two years, managers can assess the stability of the funding and thereby also assess risk. While long-term funding is not guaranteed, project managers should consider its impact on programs and projects—and in doing so, assess long-term budget-stability risk, as well.

Program Reliance on Set Cost and Schedule

To what level does a program depend on the cost and schedule being set? For example, does the program need to hit a specific launch window for another celestial body, and is such a schedule maintained at the expense of cost? Or can managers make trades between costs and schedule?

Insource/Outsource Trade-Offs in Cost and Flexibility

Some tasks that NASA asks partners to perform it can perform internally. Its capability to do so is part of the overall risk assessment. Alternatively, as noted, NASA may choose an external partner for reasons of cost, schedule, or capacity. The risk of this choice can be assessed two ways. First, it can assess flexibility on a given choice on a scale of 1 (NASA can perform the outsourced function) to 5 (NASA cannot perform the outsourced function). Second, it can assess cost using a similar scale of 1 (least expensive) to 5 (most expensive).

Human Capital Risks

Human capital risk refers to the risks posed by having the necessary skills for a mission or program. It includes the ability to find skilled individuals.

Maintaining a skilled and knowledgeable workforce is key to NASA's continued success. Like all organizations, without personnel, NASA could not operate. NASA has a human capital planning guide for its decisions (NASA, 2015a).

The Space Shuttle program helps illustrate the extent of human capital risks. Each shuttle carried seven astronauts into orbit, but the number of personnel involved with each launch numbered in the tens of thousands. That is, each shuttle launch required an extensive amount of human capital. Through the program, mission planners needed to ensure that the proper workforce was available, trained, and had the skills necessary for support. At the completion of the program, each organization supporting the shuttle reduced its workforce. Altogether in Brevard County, Florida, 9,000 workers directly supporting the space program lost their jobs; a single NASA contractor, Boeing, alone cut 510 workers (Suciu, 2011).

Osburg et al. provides further insight into the human capital risks faced by NASA:

The flight research workforce is the hardest capability to replace (if, for example, it had to be reestablished after a temporary reduction)—recruiting and maturing skilled flight research staff takes longer than adding instrumentation and other modifications to a new aircraft. The following staff skills and experience are unique to flight research:

- researchers
- airworthiness and flight safety staff (particularly for NASA-unique safety standards and procedures)
- experimental test pilots, who need R&D [research and development] qualifications (which go beyond T&E [testing and evaluation] qualifications), including for formation flight
- support engineers and technicians.

Maintaining a balanced workforce is challenging not only because of budget concerns but also surges in flight research campaigns. Although gaps might be met using a contracted workforce, there may be issues of skills availability to consider with this option.

For efficient and safe execution of flight research, all of these specialists need to work together closely and contribute to each other's activities and decisionmaking processes. Thus, collocation (or at least the availability of tools and processes for effective remote collaboration) is important, which limits the amount of outsourcing that is advisable and makes geographically distributed operations less efficient.

Based on our discussions with flight research managers at the four NASA centers involved in flight research, the workforce is aging at the same time that employment caps and lags makes hiring replacements difficult, if not impossible. At some centers (but not across NASA), the depth of expertise is already "very thin (one deep)" in some areas relevant to flight research, making it hard to compensate for illness and other unexpected events and to train replacements. As one SME [subject-matter expert] pointed out, staff members involved in flight research also need to have the right mindset: results-oriented rather than procedure-focused. This can be expanded to apply to the "corporate culture" of NASA and its centers. In this context, the workforce management challenges mentioned also make it more difficult to implement organizational and cultural change through changes in workforce.

Finally, breakthrough efforts and successes also attract more highly skilled staff, leading to more breakthroughs—a virtuous cycle, but one that can turn vicious in the absence of breakthroughs. (Osburg et al., 2016, p. 42)

Human capital risk assessments need to include several areas, including civil servants, contractor support services, demographic issues, training, workforce skill mix, work quality, and adaptability to changing missions.

Human Capital Risk Components

Human capital risks can be characterized using a number of components and metrics. Some of these metrics and indicators can be quantified, while others may only be characterized qualitatively. Components include:

- technical expertise
- availability of talent
- age of talent
- cost of talent
- adaptable skill mix/adaptability to changing missions
- training programs in place.

Technical Expertise

NASA needs skilled individuals who understand and have responsibility for all technology, with support pyramids within NASA and industry. An example support pyramid is shown in Figure 3.2.

The pyramid shows the support system for experts (at the top) down to early career talent at the bottom. The talent pyramid supports the responsible individual, as well as shows where the talent pool is currently, giving an idea of where it needs to grow. This need also pertains to technical risk. Does NASA have the technical expertise necessary to ensure technical success of a mission or missions?

Availability of Talent

Having the necessary talent employed is not the same as having that talent available for tasking. NASA employees work across multiple projects and their available time is

Figure 3.2
Example of a Support Pyramid

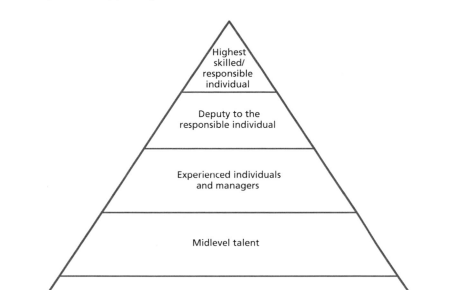

RAND *RR1537-3.2*

finite. Is there a labor base available to provide NASA with the skills it needs? Does the skilled labor have experience in the correct areas? Are they in the correct location?

Age of Talent

Understanding the age structure of the workforce can help in determining whether it will be available when needed. If an individual possessing a skill needed in ten years plans to retire in five, then there is risk to that skill availability.

Cost of Talent

The workforce associated with programs/projects (i.e., the "standing army") is a resource. How large is the workforce associated with a project or program? If the schedule is extended, how much cost is there to maintaining the needed workforce—including replacing attriting workers if necessary?

Adaptable Skill Mix/Adaptability to Changing Missions

In the private sector, workforces are actively shaped to reflect work demand. Federal government workforces are not as resilient due to federal government employment rules. Yet the NASA workforce needs to be able to adapt to changes in NASA's portfolio of work.

Training Programs in Place

Training programs are necessary to develop a capable workforce. Are training programs in place to train new workers? If so, then there may be a low level of risk associated with this component. If there are no programs in place, or programs are inadequate, the resulting human capital risk will be higher.

Organizational and Managerial Risks

Organizational risk is the risk to NASA from having a distributed workforce across multiple diverse locations with different levels of management. This risk encompasses the strength of leadership, congressional interest, and funding dispersion. Leadership and culture is different at each center.

The current organization of NASA is the result of decades of development. NASA HQ in Washington, D.C., is supported by a number of centers across the United States. These include:

- Ames Research Center in Mountain View, California
- Armstrong Flight Research Center in Edwards, California
- Glenn Research Center in Cleveland, Ohio
- Goddard Space Flight Center (GSFC) in Greenbelt, Maryland
- Jet Propulsion Laboratory (JPL) in Pasadena, California[1]
- Johnson Space Center (JSC) in Houston, Texas
- Kennedy Space Center (KSC) in Cape Canaveral, Florida
- Langley Research Center in Hampton, Virginia
- Marshall Space Flight Center (MSFC) in Huntsville, Alabama
- Stennis Space Center in Hancock County, Mississippi.

In addition to these NASA centers, a number of other facilities support current NASA missions. These include the Michoud Assembly Facility, White Sands Test Facility (managed by JSC), Vandenberg Air Force Base (managed by KSC), Wallops Flight Facility (managed by GSFC), and the Software Independent Verification and Validation Facility (managed by GSFC). All of these facilities are shown in Figure 3.3.

Having managers, researchers, technicians, facilities, equipment, and manpower spread across the country poses organizational risk to NASA. The geographical challenges faced by NASA as it develops and performs work are a risk as well. Skills needed for a given function may be spread across multiple locations spanning multiple time zones. Cultural differences among NASA sites can also lead to different approaches in performing work.

---

[1] JPL is a federally funded research and development center.

Figure 3.3
NASA Center Geographical Locations

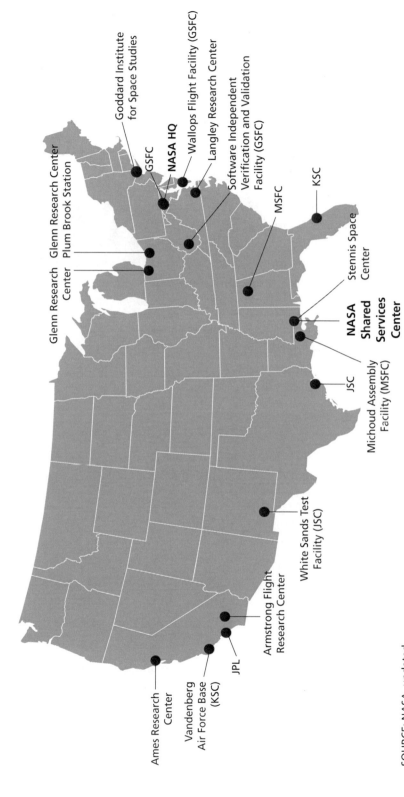

SOURCE: NASA, undated.
RAND RR1537-3.3

Organizational Risk Components

Organizational risks can be characterized using a number of components and metrics. Some of these metrics and indicators can be quantified, while others may only be characterized qualitatively. Components include:

- strength and interest of leadership
- levels of management involved in work
- number of locations involved in work
- dispersed management of projects and funds
- congressional backing of individual locations
- cultural differences between locations.

Strength and Interest of Leadership

Leadership strength affects an organization. Leaders give organizations direction, set expectations, and provide an interface to the public. The level of interest that leadership shows in a program, project, or technology can affect organizational risk as well.

Levels of Management Involved in Work

Answering to multiple levels of management can be time-consuming. Some projects, however, involve multiple levels of management. Individual employees need to navigate these and resolve potentially contradictory directives, resulting in delays coming from different levels.

Number of Locations Involved in Work

More locations can increase the complexity of a project and therefore the risk. However, additional locations can also help a project gain funding. Other RAND work (Osburg, 2016) has shown that the centers collaborate but are not unified, and that HQ does not have much power to enforce change among the centers.

Dispersed Management of Projects and Funds

Funding to projects from multiple sources, or from a single source but to multiple work locations, adds complexity and risk. Managers can have different work styles and expectations. When managers work across different locations, directorates, or even in the same location but on different projects, these styles and expectations can be a source of friction and therefore risk.

Congressional Backing of Individual Locations

Each of the centers is in a location with different congressional support. Some members of Congress are seasoned lawmakers successful at getting legislation passed, while others are less experienced. Their support to NASA, including the time they spend at or working with a NASA center, varies.

Cultural Differences Between Locations

Differing work cultures and styles can also pose an organizational risk. Such risk differs from that posed by the number of work locations and relates more to the work style that each location has. Some NASA centers may be closely aligned culturally to each other; others may be quite different.

External Dependency Risks

External dependency risk is the risk to NASA posed by reliance on other organizations with different strategies and stability.

NASA relies on other U.S. government and nongovernmental organizations for services, facilities, and support outside its own supply chain. Such partnerships can help reduce costs but do pose risks.

The Orion program helps illustrate such risks. The Orion crew vehicle will parachute into the ocean and be retrieved by the U.S. Navy, as the Navy also retrieved Apollo program capsules in the late 1960s and early 1970s. This constitutes a dependence on the Navy for manned spaceflight. Due to the size and weight of the Orion capsule, Navy ships with a well deck (a large opening in the rear of certain ships that can be flooded so equipment and ships can float in and out of the host vessel) are required for safe processing and transport of Orion capsules after splashdown (Figure 3.4).

Figure 3.4
USS *Anchorage* Recovering Orion Spacecraft in Her Well Deck

SOURCE: Herridge, 2014. Photo credited to U.S. Navy.
NOTE: The USS *Anchorage* is an amphibious transport dock ship (LPD 23).
RAND *RR1537-3.4*

The U.S. Navy currently has around 30 active ships with well decks. This number should stay roughly the same in the future. In the past decade, however, the Navy decided to build a new class of large-deck amphibious ships without well decks (Program Executive Office Ships, 2009). The *America*-class landing helicopter assault ship was meant to be the first of 11 such ships without well decks and envisioned to replace those with them. The Navy later changed its decision, stopping *America*-class manufacturing at two ships, and adopting an altered design that included well decks. For now, NASA should be able to rely on the Navy for crew and ship recovery, but the *America*-class decision shows how NASA's reliance on external organizations can be a risk. Had the *America*-class gone forward as originally planned, the number of well deck ships in the Navy would have dropped by about a third.

External Dependency Risk Components

External risks can be characterized using a number of components and metrics. Some of these metrics and indicators can be quantified, while others may only be characterized qualitatively. Components include:

- partnership and funding approaches for stability
- level of dependence
- amount of funding
- primary mission, strategy, and planning of dependent organization
- stability and strategy of dependent organization
- alternatives for dependence.

Partnership and Funding Approaches for Stability

The relationship between NASA and the external organization is one component of the overall risk assessment. Part of this component is the type of funding that NASA and the external organization receive. Is NASA being funded to perform a mission? Does the external organization draw some of the funds as NASA does, or is its funding separate? If the external organization is a private organization, NASA might have some control due to the contracting of the work and therefore could keep a stable budget as it relates to NASA. If the external organization is a government agency or department, the budget stability is likely out of NASA's control. If the external organization is a foreign government, the budget outlook could be even more risky. In this case, is the organizational relationship with NASA a partnership, as in the case of the Alternative Fuel Effects on Contrails and Cruise Emissions II (ACCESS-II) testing? Partnerships provide NASA projects and programs with the potential for higher aggregated funding and scope, but also rely on foreign governments to provide what NASA needs. Alternatively, NASA may partner with private organizations for a particular mission or program and consider the financial stability of these organizations in its risk assessment.

This component has several pieces. The first is whether funding for a program comes from NASA or elsewhere—with risk increasing to the extent that funding relies on external sources. Second is the nature of the organization, with other government agencies posing less risk than private or foreign partners. Financial stability of the external organization is the third piece of this category.

Level of Dependence

A program's level of dependence on an external organization is one component of external risk. If a mission cannot succeed without the external organization, the mission is wholly dependent on it. If the external organization plays a part but the mission is not reliant on it, then the risk to the program or mission is lower.

Amount of Funding

Different from the level of dependence, this category explores the actual amounts of money that NASA gives to the dependent organization and the percentage of its overall funding that the dependent organization receives from NASA.

Primary Mission, Strategy, and Planning of Dependent Organization

The primary mission of the dependent organization should be part of the analysis. As the Orion example shows, the U.S. Navy has a mission that, at times, can be incongruent with that of NASA. Navy ships are designed to perform warfighting duties. While the well deck of the Navy's amphibious ships is suited to recovering the Orion capsule, its primary function is to launch and recover landing craft carrying U.S. Marines and their equipment. Decisionmakers in the Navy will always choose their primary mission over the NASA mission. The Navy is, however, under the command of the President, who also oversees NASA—which can benefit NASA, should resource constraints hit the Navy.

Foreign entities that work with NASA—such as Roscomos (the Russian Space Agency), Japan Aerospace Exploration Agency (JAXA), German Aerospace Center (*Deutsches Zentrum für Luft-und Raumfahrt* or DLR), European Space Agency (ESA), National Research Council of Canada, and Netherlands Aerospace Centre (NLR)—are under the control of foreign governments. Nevertheless, their missions may align closely with that of NASA. Subject-matter expertise can help assess the mission of dependent organizations, their history of working with NASA, and other relevant characteristics.

Stability and Strategy of Dependent Organization

U.S. government agencies and departments are always at risk for budget cuts. Foreign government agencies face similar risks. Private institutions, which rely on the support of shareholders or directors, have their own rules for operations. Every organization has its own strategic plan. Working with NASA may or may not be part of the dependent organization's planning.

To assess the risk that other organizations' plans may pose to NASA, analysts must discern how closely aligned the strategic planning at the dependent organization is with NASA's goals—as well as the future stability of the organization.

Alternatives for Dependence

Regardless of the level of dependence, available alternatives can alleviate risk associated with external dependency. Having a capability available from multiple organizations can reduce organizational risk.

Political Risks

Political risk is the risk to NASA posed by changes in direction from elected leaders, changes in program budgets, the federal budget process itself, and the restrictions placed on working with international entities. These factors are largely outside the direct control or influence of NASA.

NASA's reliance on Russian aerospace companies and the government of the Russian Federation is an example of a political risk. Agreements between NASA and these organizations are controlled by the executive and legislative branches of the U.S. government. In recent years, the relationship between NASA and Roscomos has been the subject of discussion due to U.S. sanctions on Russia following Russia's actions in Crimea. NASA depends on Russian rockets and launch services to access the ISS (Kremer, 2014). Without Russian involvement, NASA would have no means to reach the orbiting laboratory. That is, NASA access to the ISS is subject to a political risk outside its control (Carney, 2015).

Political Risk Components

Political risks can be characterized using a number of qualitative components and metrics. Electorates choose political leaders who decide NASA's mission and budgets. The missions and budgets may not align. An agreement made by NASA under one set of political leaders may change if the leaders change or if the environment under which the leadership operates changes. Components include:

- changes in direction from the political leaders of the country
- reliance on foreign entities
- congressional restrictions
- federal budget process.

Changes in Direction from the Political Leaders of the Country

Shifts in policy can occur between presidential administrations. Each presidential administration brings new priorities to NASA. Often, these priorities replace previous priorities. With a presidential administration change possible every four years, and a relatively finite number of candidates, NASA could internally analyze the possible priorities.

Accounting for political risks should also account for changes in national strategy and goals. The national strategy toward space exploration, aeronautics research, and science can change, meaning existing NASA projects and programs will be affected by changes in the presidency.

Reliance on Foreign Entities

Reliance on foreign entities for services means NASA faces risk from decisions made by foreign governments and companies. NASA should account for this, as well as risks posed by possible domestic restrictions on purchases of foreign equipment.

Congressional Restrictions

Congress can specify how NASA's funds are spent. Recent congressional opposition to NASA Earth Science missions shows how changing congressional priorities place NASA plans at risk. Congress also can restrict NASA's interactions with a foreign government, thereby restricting NASA's ability to use foreign products and services.

Federal Budget Process

The federal budget process itself poses a risk to NASA. Lawmakers debate the budget largely outside of NASA's influence. Occasionally, the budget process may lead to a government shutdown and furloughs, adversely affecting NASA's work.

Technical Risks

Technology also poses a risk to NASA operations. Technical risk includes the readiness of the technology, documented understanding of the technology, safety of the technology, and reliance of programs on a given technology.

Of the risks discussed in this report, technical risk may be the one NASA traditionally assesses the most (for instance, see Weber et al., 2012 for a review of use of Bayesian networks; other examples include Frank, 1995; Paté-Cornell, 2001; Mohaghegh, Kazemi, and Mosleh, 2009; Boyer and Hamlin, 2011). As a result, NASA has guidelines for technical risk. For instance, it has a risk-informed decisionmaking handbook that describes how technical, schedule, cost, and safety risk are assessed within the agency's decisionmaking process. This process is shown in Figure 3.5.

While the figure and risk-informed decisionmaking process are designed for mission decisions, they can be used at a higher level to assess NASA technical risk for the purposes of this analysis.

Figure 3.5
NASA Risk-Informed Decisionmaking Process

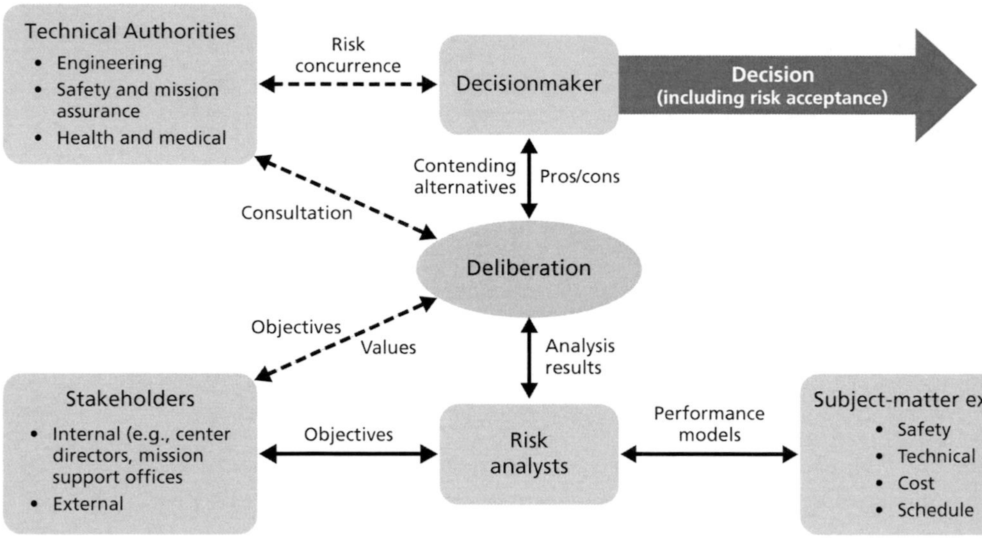

SOURCE: Dezfuli et al., 2010.
RAND *RR1537-3.5*

Technical Risk Components

Technical risks can be characterized using a number of components and metrics. Some of these metrics and indicators can be quantified, while others may only be characterized qualitatively. There are a number of technical risk components that NASA should consider and assess when developing overall risk assessments. Components include:

- documentation for key technology cost and schedule assumptions
- payoff of technology
- readiness levels (TRL, MRL, IRL)
- safety of the technology systems
- reliance on technology of other programs.

Documentation for Key Technology Cost and Schedule Assumptions

Documentation and peer reviewing of assumptions would help NASA develop more-accurate assessments.[2] This will also allow persons responsible for the key technology to assess technologies and make comparisons. A lack of documentation and understanding in the community may increase technical risk.

---

[2] A criticism of NASA is the workforce's over-optimism at developing technology risk assessments in terms of cost and schedule.

Payoff of Technology

Investing in state-of-the-art technology can help NASA make breakthroughs. Such investment should have a planned payoff. There is a risk that investing in a potential new technology will not enable new missions, or improve performance of existing missions. A system, particularly an enabling technology, that needs great amounts of resources may hold risk due to the resources being used at a single point of failure. Risk can be balanced between the necessary resources for development and what the technology will enable. High-cost and high-payoff technologies may hold high risk but be worth the investment. High-cost and low-payoff technologies holding high risk may not be worth the investment.

Readiness Levels

Understanding where a technology falls on the readiness scale can indicate the amount of risk being taken with its use. Low readiness for a planned technology can pose a high level of risk because there is no guarantee that the technology will mature to mission capability. Common numerical assessments of technology maturity are TRLs, MRLs (Homeland Security Institute, 2009), and IRLs. The TRL scale is a numerical assessment of the maturity of a given technology. The MRL scale is a numerical assessment of how close the technology is to being manufactured in industry. The IRL scale is a numerical assessment of how close the technology is to being ready to integrate with other technologies.

Safety of the Technology Systems

System safety constitutes another part of overall technology risk assessment. No exploration system is ever completely safe, so perceived safety is an important part of the overall assessment.

Reliance on Technology of Other Programs

Sometimes programs plan to use equipment and systems that were developed for another program. Hence, consideration of technology risk should consider whether a program or project assumes that a key piece of technology will be available from another program. For example, if Program A is developing a widget that Program B plans to use and Program B therefore devotes no funding to developing the same widget, then Program B faces a risk in its reliance on Program A (e.g., Program A may not develop the widget after all, or develop it late).

Additional Considerations

Many of the risks and their components reviewed here are interdependent. For example, cost and schedule risks vary with technical risks, as technology problems may lead to cost and schedule growth or variances. Similarly, organizational risks vary with technical risks. If organizational centers or facilities that are managing different components of a system are not in coordination, it is possible to introduce integration risks into the system. While the methodology we present in the next chapter does not eliminate these interdependencies, proper development of indicators and components using conditional statements will help account for them.

Personnel affect most, if not all, risks we consider. Many of the risks we identified may be affected by personnel who develop, produce, and field technologies; manage the business of an organization; train relevant workforces; and may make their own errors when managing risks. In short, analysts and decisionmakers must also consider human elements when assessing risks.

CHAPTER FOUR

Development of the Risk-Informed Decision Methodology

This chapter focuses on the development of the risk-informed decision methodology for conducting a holistic evaluation of mission or project risk that NASA assumes through various high-level decisions. Our methodology seeks to support NASA in understanding associated risks and making decisions related to certain projects or missions.

This chapter builds on the elements of Chapters Two and Three to arrive at a methodology that can provide insights into NASA organizational risk. In Chapter Two, we reviewed general literature on risk, focusing on knowledge of risks and their assessment and management. In Chapter Three, we defined seven specific risk factors—supply chain, cost and schedule, human capital, organizational and managerial, external dependency, political, and technical risks—and identified components, indicators, and mitigation methods for each of these. With this foundation, in this chapter we will incorporate the body of knowledge concerning risk and decision assessments with the specific definitions of individual risk factors to develop a methodology for examining NASA organizational risk.

This RAND-developed risk-informed decision methodology uses a high-level assessment fit on top of lower-level assessments that allow for analyzing individual risk elements. The subordinate analyses will allow for normalizing and assessing the NASA-level risk associated with a particular decision or set of decisions. As such, the methodology will provide a basis for developing a NASA-level risk-informed decision analysis that facilitates comparing, understanding, and mitigating risks likely to result from organizational-level decisions.

Figure 4.1 depicts the methodology for developing an organizational-level assessment for a particular decision or set of decisions. It uses a summation sign to illustrate the combination of the seven risk assessment factors. This is not to imply that the factors are summed to arrive at a total value for risk assessment. Rather, it demonstrates the terms are combined within an overarching risk-informed decision analysis.

The culmination of this combination of disparate risk assessment factors will be the display of the risk for a particular NASA issue on a radar chart, with each spoke corresponding to a risk factor. The most important aspect of such a display is the normalization of the risk factors (i.e., converting all risk factors to the same scale) so that comparisons can be made.

Figure 4.1
Assessing Organizational Risk

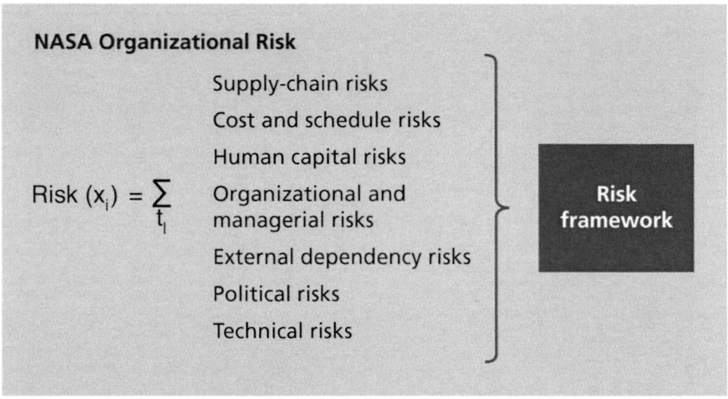

RAND RR1537-4.1

Developing the Methodology

Normalizing risk factors requires having common units in terms that can be compared with one another so that, for example, the extent of political risks can be compared with that of supply-chain risks. (See Chapter Two's discussion on multi-attribute methods.)

In such an analysis, the absolute values are far less important than the comparison of the values.[1] Allowing such a comparison is not to say that one can trade technical risk mitigation for political risk. Rather, such normalization allows one to depict where NASA leadership should have the greatest concerns and ultimately where resources may be allocated to mitigate risks and improve chances of mission success.

The centerpiece of the normalization process is the analysis of risks and their components. Figure 4.2 depicts the seven risk factors and 37 components that we identified. Each component must be analyzed to identify the metrics associated with the risk factor, the methods to evaluate the risks, and the measures that may mitigate them.

Components must also include the boundary conditions that define the acceptable limits of each component. Ranges with maximum and minimum limits must be established to serve as these boundary conditions, where relevant. In analysis of each component, values or assessments that exceed the boundary conditions are by definition unacceptable and therefore cause for either rejecting the entire program under consideration or for incorporating mitigation measures. The incorporation of mitigation measures must be done to ensure that all boundary conditions are eventually met. Should a component boundary condition not be met, the overall risk is determined to be unacceptable for the entire risk factor being analyzed.

---

[1]   In the classical risk assessment format, when risks are represented by the likelihood of some outcome occurring (e.g., 10 percent of five fatalities), absolute values are indeed very important. Once these risk values have been normalized to a common ordinal scale (e.g., 1 to 5), the values become less descriptive and therefore lack the actionable aspect of their classical counterpart.

Figure 4.2
Risk Factors and Components

Supply chain	Cost and schedule	Human capital	Organizational and Managerial	External dependency	Political	Technical
Availability of materials	Cost for maintaining access to needed resources	Technical expertise	Strength and interest of leadership	Partnership and funding approaches for stability	Changes in direction from political leaders	Documentation for key technology cost and schedule assumptions
Availability of services	Schedule for maintaining access to needed resources	Availability of talent	Levels of management involved in work	Level of dependence	Reliance on foreign entities	Payoff of technology
Stability of sources	Budget stability	Age of talent	Number of locations involved in work	Amount of funding	Congressional restrictions	Readiness levels (TRL, MRL, IRL)
Availability of alternative sources	Program reliance on set cost and schedule	Cost of talent	Dispersed management of projects and funds	Primary mission of dependent organization	Federal budget process	Safety of the technology systems
Quality management	Insource/outsource trade-offs in cost and flexibility	Adaptable skill mix/adaptability to changing missions	Congressional backing of individual locations	Stability and strategy of dependent organization		Reliance on technology of other programs
		Training programs in place	Cultural differences between locations	Alternatives for dependence		

■ Risk factor
□ Component

For each component, must identify:
- Metrics
  - Boundary conditions (mins and maxs)
  - Ranges
- Evaluation methods
  - Qualitative
    - Expert elicitation?
    - Literature review?
    - Other?
  - Quantitative
    - Flow models?
    - EVM?
    - Other?
- Mitigation
- Normalized values tied to ratings:
  - Unacceptable
  - Extremely high
  - High
  - Moderate
  - Some
  - Low

RAND RR1537-4.2

Evaluation methods should be identified for each component. These will vary by risk factor and component. Some will lend themselves to quantitative measures, while others will be analyzed qualitatively. A mix of objective and subjective measures will likely be employed to evaluate each component. A number of these methods are presented in Chapter Two, including EVM and Balanced Scorecard methods.

These elements of the methodology should be considered in each risk assessment but the definitions of each, boundary conditions, thresholds for normalization, and mitigation strategies will vary based on the case under consideration. For example, in the cases we considered, the previous decision to cancel the Space Shuttle program and commercialize transport to the ISS and the future decision on the Cislunar Habitat, the methodology remained constant but the data elements changed.

This is not to suggest that this list of factors is complete and could not be modified. One might want to see a stronger representation of such areas as legal, reputational, or infrastructure risks, for example. Such additions could be inserted, either as new risks or risk components, with no loss of fidelity in the methodology. Such additions would require developing documentation on definitions, boundary conditions, and evaluation methods for the new topics to be considered.

The methodology allows converting the information—whether it is quantitative or qualitative—to a common scale, using an agreed set of threshold terms. When agreed values are associated with the different risks and their components, comparison among them is possible. We discuss such normalization further below.

The Normalization Process

The structure for the normalization process begins with the identification of six levels of risks, ranging from low to unacceptable, in which higher numbers indicate greater risk to the project or mission. Here, low risk corresponds to 0 while unacceptable risk corresponds to 5 on the scale. Figure 4.3 depicts the rating scheme applied in the normalization process. Once the data for each risk are analyzed and an assessment as to the relative risk level is made, the point can be plotted on the radar chart and the relative risk can be displayed.

In such a process, any risk where one of the component boundary conditions is violated will, by definition, be assessed to be in the unacceptable range unless it is mitigated to an acceptable level, resulting in an adjustment to the risk factor rating. In this case, an unacceptable rating would indicate too great a risk to mission or project success. A discussion of risk acceptability can be found in Chapter Two.

Understanding the risk factor normalization rating system is only the initial part of the analysis. The foundation of the system begins with defining the risk factors and components such that a mix of analytical methods can be applied to understand the risk associated with each.

Figure 4.3
Assessing External Dependency Risk—Normalization

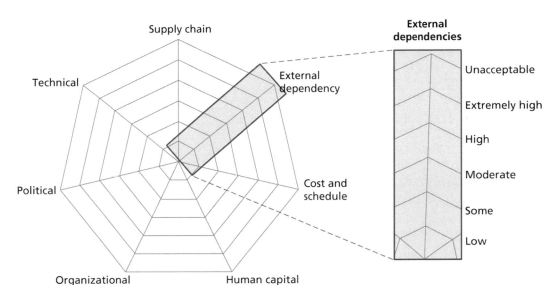

RAND *RR1537-4.3*

For each risk, the components are defined, the indicators identified, evaluation methods presented, and mitigation possibilities identified. Figure 4.4 presents an example of this process using ISS transport. This information will serve as a basis for conducting the analysis that ultimately will result in a normalized value being identified for the risk factor—in this case, the ISS transport supply chain.

The radar chart and supporting documentation will provide a basis for the development of a NASA-level risk-informed decision analysis that allows for comparing, understanding, and mitigating risks likely to result from organizational-level decisions.

This initial analysis is designed to develop a basic understanding of the relative contribution of each risk in assessing the overall risk of the issue (like a project or mission) under consideration. It also allows for assessing whether a more cursory or intensive analysis of the risk and its components is warranted. If detailed information is not available, then a more cursory review may be appropriate. One might also choose a more cursory analysis if the initial evaluation indicates that the risk is unlikely to significantly affect any risk-informed decision analysis.

Regardless of whether a cursory or more thorough review is employed, and regardless of the tools used in assessing risk, the values assigned to the risk can be normalized and plotted on the radar chart. In fact, the methodology allows the plots on the radar chart to contain a mix of values from cursory or more-exhaustive examinations. If the cursory methodology is employed initially and additional information later becomes available, or if the importance of a risk grows, one can recalculate the value and replot the result. We discuss techniques for doing so.

Figure 4.4
Supply-Chain Assessment Information for Cancellation of the Space Shuttle and Commercialization of Transport to the ISS

Supply Chain	Indicators	Evaluation Methods	Mitigation
Stability of sources for components and equipment			
Ability of commercial sector to provide stable supplies to the ISS in a timely manner	• Cost/schedule delays • Number of vendors • Performance • Reliability • Number of sources • Acceptable levels	Critical Path Methods; Qualitative	Accelerating readiness of supplies by investing more funds earlier; diversification of commercial partners; deferring retirement of Space Shuttle program
Availability of alternative sources (domestic and foreign) for components and equipment			
Alternative supply-chain sources and dependence on foreign countries	Alternative supply-chain sources available during each time period	Number of alternative supply-chain units required over time	Same as above
Availability of materials			
Availability of necessary materials	Units of material	Number of units available	Divide among options identified above
Availability of services			
Commercial sector to provide necessary services to the ISS	Units of services required per unit time	Number of units available	Divide among options identified above
Quality management			
Quality of the product delivered	Comparison to baseline quality	Quality assessment by subject matter	Rigorous quality management programs

RAND *RR1537-4.4*

The Detailed Assessment—Intensive Methodology

The detailed assessment relies on a comprehensive assessment of the components and their indicators for each risk. As previously explained, our methodology involves evaluating a number of *risk indicators*, or knowable metrics that describe different aspects of the current state of the world that provide an *indication* of the risk (i.e., the likelihood[s] that certain consequence[s] of concern to that organization will occur). Such a methodology may be appropriate when probabilities of future scenarios are unknowable.

Characteristics of such an assessment are likely to include reliance on analysis using the full gambit of tools ranging from cost, schedule, and performance methods such as EVM to modeling and simulation (where available) to assess a component in detail. Input for such tools will likely rely on subject-matter expert interviews, data elicitation, and literature reviews. Figure 4.5 depicts a worksheet for capturing the results of the analysis and calculating the unmitigated and mitigated normalized risk values

Figure 4.5
Detailed Assessment Worksheet for Stability-of-Source Component for Supply Chain

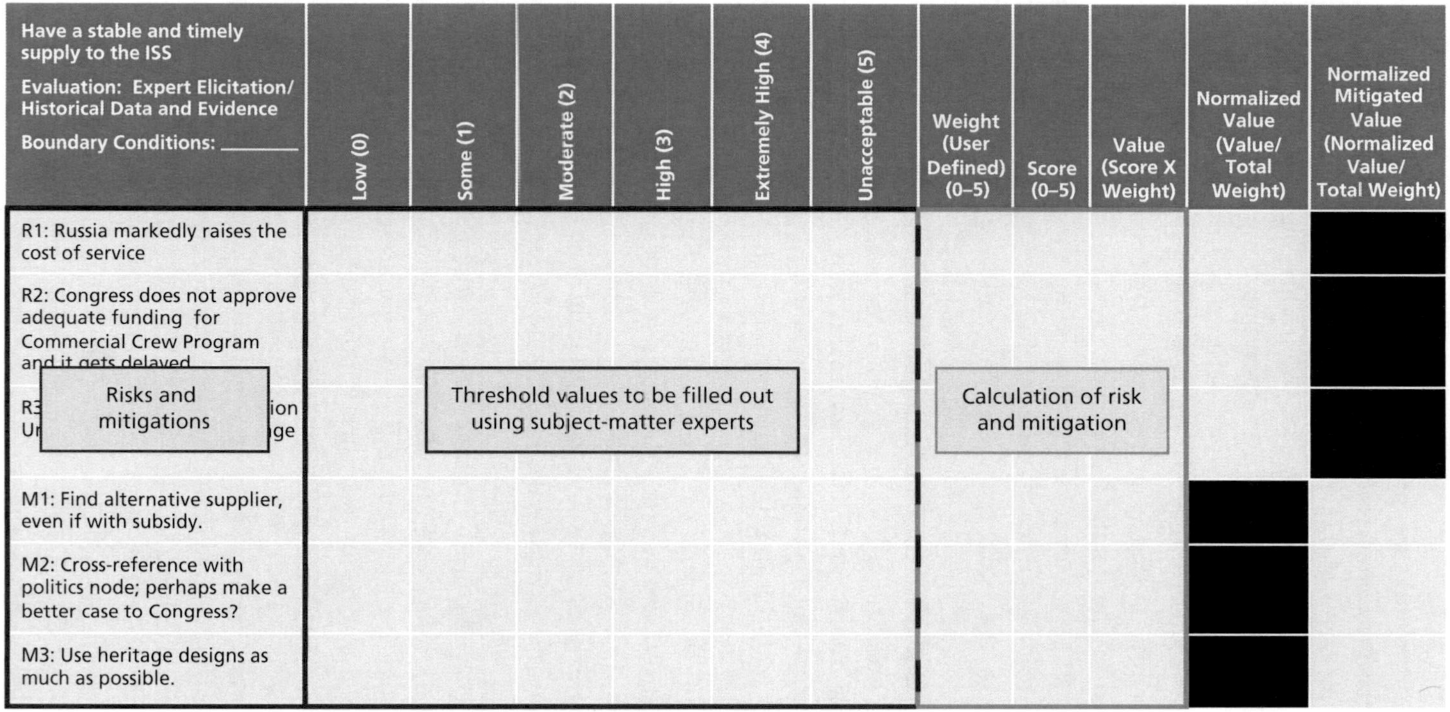

Have a stable and timely supply to the ISS Evaluation: Expert Elicitation/ Historical Data and Evidence Boundary Conditions: _____	Low (0)	Some (1)	Moderate (2)	High (3)	Extremely High (4)	Unacceptable (5)	Weight (User Defined) (0–5)	Score (0–5)	Value (Score X Weight)	Normalized Value (Value/ Total Weight)	Normalized Mitigated Value (Normalized Value/ Total Weight)
R1: Russia markedly raises the cost of service											
R2: Congress does not approve adequate funding for Commercial Crew Program and it gets delayed											
R3: ...											
M1: Find alternative supplier, even if with subsidy.											
M2: Cross-reference with politics node; perhaps make a better case to Congress?											
M3: Use heritage designs as much as possible.											

Risks and mitigations

Threshold values to be filled out using subject-matter experts

Calculation of risk and mitigation

RAND RR1537-4.5

for a component. These normalized risk values can be summed to provide an overall value for each risk factor.

The detailed assessment worksheet provides considerable information for assessment-team leaders. The left column contains information on the risk factor component. The top block defines the component of risk considered, the evaluation methods employed in assessing it, and the boundary conditions for it. Also in this column, bordered in purple, are individual risk indicators (e.g., R1) and corresponding mitigations (e.g., M1). Listing indicators and mitigations separately allows calculation of unmitigated and mitigated risk that will help NASA leaders understand the risk arising from particular courses of action. The risk-assessment team leaders are responsible for developing this information.

The second section of the detailed worksheet, bordered in red, would be completed with input from subject-matter experts estimating the threshold values for the risk levels shown in the top row, for each indicator and its mitigation. These values are critical as the normalization process essentially relies on them. Having senior NASA personnel assist in developing these values is essential. Because risks will have inter-dependencies, indicators of their components and determinations of their threshold values should be made conditionally. For example, cost, schedule, and performance indicators are necessarily intertwined. A decisionmaker may decide that obtaining performance within 95 percent of the objective is paramount. Thus, cost and schedule thresholds should be set with this in mind, understanding that such a high standard on performance could mean that cost or schedule growth would be acceptable.

The third section, in the green box, contains the risk and mitigation calculations. The first column is a user-defined weight ranging from 0 to 5, where noninteger values (e.g., 1.5) are permitted. This allows for increasing the relative importance of particular risks or mitigations in the analysis and is reflected in the component calculation of risk. The second column translates the indicator value calculated in the analysis using the threshold values from the second section to obtain a normalized value from 0 (low) to 5 (unacceptable). The third column in this section is the product of the weight and score. The values of this product may then be normalized in the fourth column. This normalization involves dividing the product of the weight and score by the total sum of the weights. These normalized values may then be summed to produce an unmitigated normalized value for the risk component. Similarly, once values are filled in for the mitigation rows, a total mitigated value may also be calculated. In this case, a mitigation will be assigned to an indicator (e.g., M1 is assigned to R1) and will reduce its overall score. Thus, if R1 had a score of 4, M1 may result in a mitigated score of 3. In some cases, two mitigations may be assigned to one indicator (e.g., M1a and M1b are assigned to R1). In this case, the mitigated score for R1 would take both mitigations into account.

This sum-product calculation is modeled after the multi-attribute methods described in Chapter Two. Note that the weights, ranging on a scale from 0 to 5, provide for an interval scale (as opposed to an ordinal one) such that an indicator designated with a weight of 2 is considered to be twice as important as an indicator with a weight of 1. Similarly, an indicator with a weight of 3 is 1.5 times more important than one with a weight of 2. Noninteger values can be used to represent fractional increases of importance (e.g., an indicator with a weight of 1.5 is 50 percent more important than one with a weight of 1). This distinction is worth considering before choosing weights. Indeed, the choice of weights can be susceptible to bias from decisionmakers. As discussed in Chapter Two, a number of methods for choosing weights are available that can reduce that bias. The method chosen for the analysis here provides one that is simple to implement and requires the least steps. As a default, all weights are set to 1. A decisionmaker should have justifiable rationale for increasing any weights.

The last two columns provide the unmitigated and mitigated risk for each component. Each is divided by half the total weight. In making these calculations, one can choose to either apply mitigation or not. If not mitigating a risk, the value is the same as for the unmitigated risk.

After the analysis has been completed, the final worksheet (see Figure 4.6) can be completed. This identifies any boundary conditions that exist and notes the overall normalized score for each risk component calculated previously on the detailed assessment worksheet (Figure 4.5). The methodology allows users to define weights from 0 to 5 so as to differentiate among components. The product of the weight and the value is recorded in the fifth column and then normalized in the last column by dividing the sum of the weighted values for each component by the total weight. This value can be plotted on the radar chart and has been normalized to reflect the risk tolerance associated with each of the seven risk factors.

Figure 4.6
Risk Factor Consolidated Assessment and Radar Chart

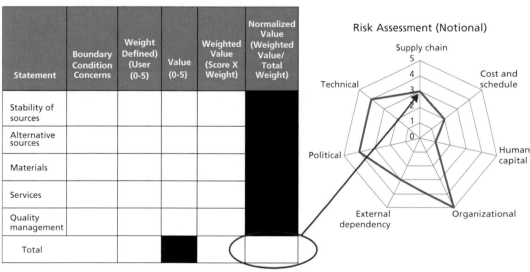

The Overview Assessment—Cursory Methodology

The overview assessment follows a similar methodology, albeit with some differences. First, it relies on a much less detailed assessment to arrive at a value for each of the risks. Instead of considering each risk component separately, it considers a relatively high-level set of statements that allow for discerning the risk associated with a particular risk factor. Figure 4.7 provides an example of such an analysis for the external dependency risk factor. Second, this methodology does not expressly consider mitigation. One could add mitigation to such an analysis, but we did not develop one for this example using only very high-level assessment criteria.

Note that NASA decisionmakers must still identify threshold values corresponding to the normalization criteria. So while this cursory method allows for a more rapid analysis of risks, conducting interviews with subject-matter experts and gaining insights into risk tolerance and threshold values would still have great utility.

The normalized value calculated at the bottom of the chart would be plotted on the radar chart as with the more detailed methodology. An analysis—and, therefore, a spreadsheet—would be developed for each of the seven risk factors, rather than for the 37 individual risk components.

Interpreting the Results

A radar chart represents the "final" results from the use of the RAND risk-informed decision analysis. This graphical depiction is likely the single most prominent depiction of the results of the analysis and the most likely to be shared with decisionmakers. Given the importance of the radar chart, several comments are in order.

First, while the radar chart is what decisionmakers will likely see, the application of the methodology is actually the most important aspect of the analysis. The development of the risks and mitigations, their threshold values, and the normalization process will provide important insights into the risk associated with the decisions under consideration. It will also offer lessons on the risk burden for different strategies and help identify the tolerance of leadership for these burdens.

Second, the radar charts depict the relative risk from low to unacceptable for each of the risk factors. Through a normalization process, one can assess whether one risk factor is of more concern than another. Absolute values associated with particular risk factors should be avoided. One can also display several different analyses on a single radar chart plot. For example, if several different strategies for a particular NASA issue were considered, one could run the analysis multiple times, employing different threshold values for each strategy, to arrive at a different perception of risk. In this way, one could graphically compare the different strategies to better understand their effect on risk.

The radar chart can also depict a single strategy and the confidence intervals that define each of the risk factors. For example, if great uncertainty surrounds the organi-

Figure 4.7
External Dependency Risk Factor Overview Assessment

External Dependency Risk Component Factor	Low (0)	Some (1)	Moderate (2)	High (3)	Extremely High (4)	Unacceptable (5)	Score (0-5)	Weight (User Defined) (0-5)	Value (Score X Weight)	Normalized Value (Value/ Weight)
Partnership and funding approaches for stability		5% <adequate	10% <adequate	15% <adequate	20% <adequate	25% <adequate	2	4	8	8/25
Level of dependence		5% <adequate	10% <adequate	15% <adequate	20% <adequate	25% <adequate	3	3	9	9/25
Amount of funding	Adequate	5% <adequate	10% <adequate	15% <adequate	20% <adequate	25% <adequate	3	5	15	15/25
Primary missions of dependent organization	Adequate	5% <adequate	10% <adequate	15% <adequate	20% <adequate	25% <adequate	4	5	20	20/25
Stability and strategy of dependent organization	Adequate	5% <adequate	10% <adequate	15% <adequate	20% <adequate	25% <adequate	4	4	16	16/25
Alternatives for dependence	Adequate	5% <adequate	10% <adequate	15% <adequate	20% <adequate	25% <adequate	4	4	16	16/25
						Total		25		3.36

NOTIONAL DATA ONLY

RAND RR1537-4.7

zational risk factor, one can reflect this uncertainty by changing the associated threshold values and replotting the results. Doing so parametrically would result in a series of confidence intervals. This approach could be taken for a single risk factor or for any combination of risk factors and would be highly appropriate for risk factors for which data are not readily available or that have great variability.

Third, given the normalization process involved in the analysis, one can actually sum the area under the lines attaching the charted risk levels in the radar chart when considering multiple strategies and make comparisons and generalizations about the overall risk associated with different strategies for decisions under consideration. In such a calculation, a smaller area would be a potential strategy that has less perceived risk, while the larger number would have greater perceived risk. The actual numbers are less important than the potential for comparing strategies.[2] For example, if risk for one strategy has an area of 25 normalized units and another has an area of 100, one can conclude that the second has greater perceived risk. That said, a calculation of 25 normalized units for one strategy and 27 for another should likely be considered as depicting equivalent risk.

Fourth, the radar chart depictions are highly dependent on the assumptions and threshold values for the strategies under consideration. In conducting a risk assessment, one must ensure that the assumptions remain valid throughout the process. Should the assumptions be found invalid or need to be changed, changes must be reflected in the threshold values and risk factor assessments. Failure to make such changes may result in an invalid risk assessment that, if followed, could lead to negative outcomes.

Methodology Caveats

The methodology just described requires careful application. Next, we note caveats. The methodology provides a *robust decision support* tool, which should not be misconstrued as a *decisionmaking* tool. This distinction is critical. To use this methodology in a rote manner and slavishly apply its result would be to misuse it. Rather, an appropriate use of the methodology is to better understand the risk tolerance of NASA leadership, develop distinct choices for consideration, and describe their inherent risks and benefits. In this way, given that NASA leadership are transparent with their preferences and risk tolerance, clear, unambiguous choices and risks will emerge that will support decisionmakers and the decisionmaking process.

---

[2] The methodology primarily uses ordinal scales, but applies mathematical operations (e.g., calculating the area of the radar chart) that are more appropriate for interval or ratio scales. This will have implications on the interpretability of results.

Risk-Informed Decision Methodology Uses and Adaptation

This methodology can be easily adapted for a variety of risk assessments. The risk assessment factors and their associated components represent a comprehensive listing of areas for evaluation. They are suitable for a broad range of cases that could be considered—as might be additional risk factors or components.

Including additional factors or components would be a straightforward process and would not adversely affect the methodological process. Should NASA desire to include an additional risk factor, it would need to develop a new spreadsheet, as well as components and their definitions and threshold values for analysis. The final radar chart depicting the total risk burden would also be expanded to include the new risk factor.

So while the basic methodology can be employed for different cases, the metrics, mitigation factors, boundary conditions, and threshold values within the methodology must be adapted for each case considered. The need for proper definitions requires that definitions be developed prior to the start of any new risk analysis.

Using expert elicitation in defining the various input values is essential. The team developing the risk indicators and mitigation strategies serves as one set of experts, while senior-level NASA personnel would be used for setting threshold values. The final sets of experts to be consulted are the decisionmakers who interpret and provide weights for the various inputs.

These groups of experts—risk team, senior NASA personnel, and decisionmakers—should be separate, lest undesirable biases or "gaming" of results be introduced into the analyses. Other requirements for conducting proper expert elicitation include ensuring that those providing inputs have the necessary expertise to provide informed answers, that experts can answer questions posed using their knowledge and the background materials available to them, and that the process uses an unbiased facilitator to structure the elicitation.

The Math Behind the Methodology

While the calculations for the higher-level assessments of the methodology are straightforward, one must be cautious in using their outputs in unintended ways. For example, simply summing values across risk factors will not provide a useful result in an absolute sense.

The results of the calculations should be to compare the burden inherent in risk factors and components, but must not be used to compare two risk calculations. For example, if technical risk is calculated as 4.5, and political risk as 4.4, both with maximum value of 5, it would be inappropriate to say that technical risk is 0.1 greater than political. Rather, technical and political risk should be considered approximately equal and both between extremely high and unacceptable based on the normalized scale.

Normalization is the Key to the Methodology

Being able to compare disparate categories of risk elements (i.e., factors and indicators) relies on a normalization process that maps these categories to a common scale. Such a process allows comparisons between qualitative and quantitative factors and displaying them in a common manner, as on a single radar chart.

Normalization requires that inputs and threshold values have equivalent "exchange rates" such that values corresponding to "high" for the external dependencies and supply-chain risk factors have a common risk perception and tolerance associated with them. In other words, leaders seeing similar values on the risk factor spreadsheets should perceive them as depicting approximately equal levels of risk. Again, the normalized values are useful for comparison purposes but not to be considered in absolute terms.

Understanding "Unacceptable" Risks

The methodology ultimately requires dividing risk into two overarching categories, acceptable (values 0 through 4 in Figures 4.5 and 4.7) and unacceptable (value 5) levels of risk. The modifiers of low to extremely high have been added to the acceptable risk category to provide a way to characterize the risk associated with a particular assessment. The term "unacceptable" serves as a boundary condition indicating that the risk burden must be mitigated or a different strategy devised.

Unacceptable risk must be considered as nonlinear. One could assess that six of the seven factors are low risk and thus receive a value of 0, yet the seventh factor could be an unacceptable risk and receive a score of 5 in the evaluation. Averaging these terms would yield a score of 0.71, which would mistakenly place the overall program under consideration in the range of low to some risk. Such a conclusion would fail to account for the unacceptably high risk in one factor and therefore the overall likelihood of catastrophic failure associated with this risk profile. Until the seventh risk factor is mitigated, the overall risk of the strategy under consideration must be considered unacceptable.[3]

Interdependencies Are Inherent

As noted, the seven risk factors and many of their components will be interdependent. Eliminating all interdependencies would require a reductionist methodology exponentially expanding the number of factors and components to be considered. Such expansion likely would not improve fidelity of the assessment sufficiently to warrant such an exhaustive analysis. Eventually, in summarizing overall risk, the large number of factors and components would need to be recombined, increasing the complexities and interactions in the final risk analysis. Therefore, the goal of the analysis was to elimi-

---

[3]   One implication of this nonlinearity is that a sum-product calculation for the total normalized value may not always be an appropriate calculation. Instead, if any component is deemed to be in the unacceptable range, the value to be used in the radar chart should be the worst normalized value.

nate interdependencies that were easily untangled and understand other areas where they existed and could not easily be eliminated.

Uncertainties

Uncertainty is a constant in complex risk assessments. The ultimate goal in preparing the risk-informed decision assessment is to describe uncertainty so leaders can better understand it and make decisions in the face of it. The assessment identifies and communicates risk factor components and indicators and their associated mitigation strategies. This process includes identifying the variables that influence and determine the values of these components and indicators. These variables include the assumptions that may change as leadership changes or new information is recognized, as well as events that could influence the likelihood of negative outcomes.

Using observable data and statistics may reduce the uncertainty that accompanies the use of subjective data from expert elicitations. These objective methods must be augmented using subject-matter expertise and elicitation to arrive at the measure for each component. As additional information becomes available, the component evaluations should be updated and the overall risk be reassessed.

Insensitivity to an Organization's Risk Tolerance

Organizations have cultures that guide their activities, the development of goals and objectives, and the manner in which risk and perhaps even failure are viewed. Therefore, understanding an organization's risk tolerance is a critical aspect of conducting risk-informed decision support and must be included the analysis.

In working to define an organization's risk tolerance, the worksheets for both the cursory and more detailed analyses will serve as useful mechanisms for detailed discussions on this topic. Such discussions will allow better understanding of boundary conditions for the various risk factors and their individual components. In some regards, these discussions become the most important elements in a risk analysis, causing leaders to acknowledge what actually is an unacceptable outcome and what is the risk tolerance their organization can bear.

Making Comparisons

The fact that such tailoring is necessary for different cases also implies that caution is needed in comparing cases. Since each case includes its own indicators, thresholds, and (possibly) components, comparisons of the same factor, such as supply-chain risks, in separate cases must be made with caution. While this methodology is well-suited to compare options in the same case (e.g., Cislunar Habitat), differences in indicators, their thresholds for normalization, and overall components complicate comparisons between cases (e.g., Cislunar Habitat and ISS transport commercialization). Careful development of these pieces of the methodology may improve comparability. Decision-makers and leaders may choose these pieces such that a "high" supply-chain risk in one case results in the same amount of concern as a "high" in another case. Such careful development would require that the same decisionmakers and leaders assess both cases and ensure that levels of acceptable risk are comparable across cases.

Eliminating Biases

Objectivity in analysis is essential to achieving the goals of the methodology. Eliminating biases is a necessary precondition for conducting these analyses. Giving responsibility for completing the elements in the Detailed Assessment Worksheet (Figure 4.5) to decisionmakers rather than risk analysts helps eliminate bias in the risk-decision assessment. The normalization process is designed to encourage scoring without regard to the final outcome, letting the methodology drive the result.

Step-by-Step Summary of the Risk-Informed Decision Methodology

Table 4.1 summarizes the steps for the risk-informed decision methodology. While the steps have been presented in detail already, this step-by-step presentation provides a consolidated description of the methodology. The table is designed to ensure the information is highly accessible to the reader.

The methodology must be updated for every assessment that is undertaken. That is, the risk factor worksheets, risk indicators, mitigations, and all values must be reconsidered as part of any new assessment. For the case studies considered—cancellation of the Space Shuttle program and commercialization of transport to the ISS and

Table 4.1
Step-by-Step Risk-Informed Decision Methodology

Step	Title	Description
1	Identification of issue	This step describes the issue, including framing the options to be considered, time frames related to the risk assessment for the decision in question, risk assessment plan, and initial listing of the experts to be consulted.
2	Validation of risk components	An assessment is conducted to ensure that the 37 risk assessment components are comprehensive and remain relevant for the analysis. A risk factor component may be found to be irrelevant. Alternatively, an expansion of the factors considered may be necessary to capture the nuance of the issue under consideration.
3	Risk factor assessment plan development	For each risk factor, a plan is developed describing how it will be analyzed. The plan includes a short description of the risk component, the indicators associated with the risk factor, anticipated evaluation methods, and the mitigation that can be employed to affect risk. (For example, see Figure 4.4.)
4	Development of the risk factor component worksheets	For each risk factor under consideration, a worksheet is developed to identify particular information about the component. (For example, see Figure 4.5.) Note: In this summary, we assume that the intensive method will be used.

Table 4.1—Continued

Step	Title	Description
4a	Description of the risk factor	A short description of the risk factor is completed and initial evaluation methods and boundary conditions are identified.
4b	Identification of risks and mitigations	Subject-matter experts who will not have a subsequent role in the evaluation identify risks and mitigations. Having independent subject-matter experts do this eliminates a potential avenue of bias in the methodology. For each risk indicator identified, potential mitigation should be identified. For some risk indicators, no mitigation may exist. For mitigations that are identified, costs should be identified as well, so that decisionmakers can make fiscally informed decisions on managing risk across NASA.
4c	Identification of threshold values	Threshold values are essential to the normalization process. Such values allow for normalization and subsequent comparison of risk factors and their components. Expert elicitation is essential for establishing these values. These values should also be developed in isolation from those charged with "scoring" the risk indicators.
4d	Establishment of risk indicators and mitigation weights	Use of weights in the methodology ensures that, where necessary, associated mitigations can be delineated by effect on overall risk. Weights can be a vital tool but must be used judiciously so as not to bias outcomes and reinforce prejudices that may be present. Each weight could initially be set to 1 and adjusted to any real number between 0 and 5 only after thorough consideration.
4e	Decisionmaker assessment	This begins with the evaluation of the score. In some cases, the decisionmaker could delegate this responsibility to a body that would assess the values for each risk indicator and associated mitigations. Such a process would best be served using experts to brief the methodology charts and request the decisionmaker or identified body to deliberate and provide a score. This process must be repeated for each risk assessment component (i.e., up to 37 worksheets).
5	Normalizing the risk indicators	The risk factor component worksheets have been developed to calculate a normalized assessment of the unmitigated and mitigated risks for each of the 37 components.
6	Calculating risk factor normalized risk (mitigated and unmitigated)	Once all risk factor components (and their risk indicators and mitigations) have been established, the risk factor roll-up can be created. The calculated values for each risk factor component are transferred to the consolidated worksheet. If both the unmitigated and mitigated risks are to be assessed, separate worksheets will need to be employed. The diagram in Figure 5.4 in Chapter Five provides a consolidation of the values obtained from the supply-chain risk factor that have been analyzed. A similar worksheet is completed for each risk factor.
7	Plotting the risk-informed decision assessment results	The risk factor worksheets from each of the seven risk factors can be used to plot the radar chart. If unmitigated and mitigated scores were developed for each risk factor, then the radar chart will have two curves plotted. Similarly, if a base case and three options were considered for a particular assessment, four curves would be plotted. Because all values were normalized, the curves can be compared and judgments made regarding particular options. (For example, see Figure 5.6. in Chapter Five.)
8	Interpreting the data	Different cases can be plotted on similar radar charts but not together on one chart. For example, one cannot plot the case studies of the next chapter together on the same chart. The methodology can be normalized within a single area but not between areas. The analysis can be used to compare options for the same case and even various mitigation strategies. One can make ordinal decisions between options using this methodology, but should not attempt to quantify the actual intervals between them.

future Cislunar Habitat decisions—the study team did a complete reevaluation of the worksheets.

Conclusions

Throughout the development of the methodology presented in this chapter, our goal has been to provide maximum flexibility to meet the needs of NASA while remaining analytically rigorous. The methodology rests on the body of risk assessment and decision analysis literature and the defined seven risk factors. Based on these foundational elements, one can identify threshold values for various elements of risk and understand how different decisions affect perceived risk. The normalization process provides the basis for comparing disparate risks to arrive at a graphical depiction of the risk burden of particular strategies and decisions under consideration. The framework described in this chapter could be used by NASA to respond to current and future requirements for enterprise risk management, such as Office of Management and Budget Circular A-123 and its requirements for enterprise risk management (Clark, 2016).

In Chapter Five, we apply this to two NASA cases—one a previous decision, the other a future decision. Though the data are notional and the "findings" of the case studies are purely to allow the authors to test the functionality, these case studies will allow us to assess the utility of the methodology and "validate" it.

Case Studies

In this chapter, we use two case studies to validate the methodology presented in Chapter Four. The first is the previous decision to cancel the Space Shuttle program and commercialize transport to the ISS. The second is the future decision regarding the scenarios for the Cislunar Habitat. In addition, in considering Cislunar Habitat risk, we consider three options, further defined below: (1) international cooperation, (2) public-private partnerships and (3) NASA-driven.

1. Cancellation of the Space Shuttle program and commercialization of transport to the ISS
   - Time frame: 2011
   - Description: The decision was to commercialize transport to the ISS and to focus NASA efforts on building the infrastructure for human missions to Mars. One key consideration is that the knowledge, lacking in the commercial sector, for building and operating space transportation systems has been transferred from one generation of NASA engineers to the next.
   - Options: None
2. Cislunar Habitat
   - Time frame: 2020–2030
   - Description: Follow-on to the ISS, the Cislunar Space Station (CLSS) will be used as the next-generation outpost near Earth for technology demonstration and habitability in space. Its goals are to improve on the lessons learned from ISS to improve life support, crew health, extravehicular activity capabilities, radiation shielding, and other areas. Much of the mission will validate processes, procedures, technology, and operations to be used on a human mission to Mars. The CLSS will give astronauts easier access to the moon, where testing can occur on a foreign astrological body close to Earth prior to committing to a Mars mission where return to Earth is measured in months instead of days.
   - Options:
     ◦ International cooperation
     ◦ Public-private partnerships
     ◦ NASA-driven.

During the course of this research, we consulted with NASA on other cases to consider. These are listed in Appendix A.

Employing the Methodology

Given study time lines and resources, full validation with intensive analyses of all risk assessment components was prohibitive. We therefore developed an abbreviated methodology in addition to the more detailed assessment methodology. We use a mix of both in our case studies, while stressing the more detailed assessment is preferable given adequate resources.

The analysis required that worksheets be developed for each of the component risk factors and their associated indicators. We present an overview of our findings in this chapter, while including all results for the case study on cancellation of the Space Shuttle program and commercialization of transport to the ISS in Appendix B and all for the Cislunar Habitat case study in Appendix C.

Important Caveat on Methodology Validation

We designed the validation to examine the limits of the methodology, assess its overall utility, and refine the methodology. We did not conduct a full risk analysis of these issues. Specifically, we did not demonstrate the feasibility of conducting an analysis that integrates multiple risk factors and compares multiple forms of risk.

We did employ internal subject-matter experts to make our studies as realistic as possible. We did not have access to actual data associated with these decisions. Instead, we made estimates and assumptions regarding the tables that depict the component risk factor indicators (including the risks and mitigations), the threshold values, the weights, and the scores for each.

But we stress that we sought more to validate our methodology than decisions regarding ISS transport or the Cislunar Habitat. This means our results must not be considered authoritative from a substantive standpoint. As a result, all tables containing data have been marked as "notional" and should only be considered as indicative of what one would expect from using this methodology.

Cancellation of the Space Shuttle Program and Commercialization of Transport to the ISS

The ISS transport risk-informed decision assessment included the more-detailed risk factor assessments for the supply-chain and political components, and the abbreviated

Figure 5.1
Risk-Informed Decision Assessment Worksheets for Shuttle Cancellation/ISS Transport

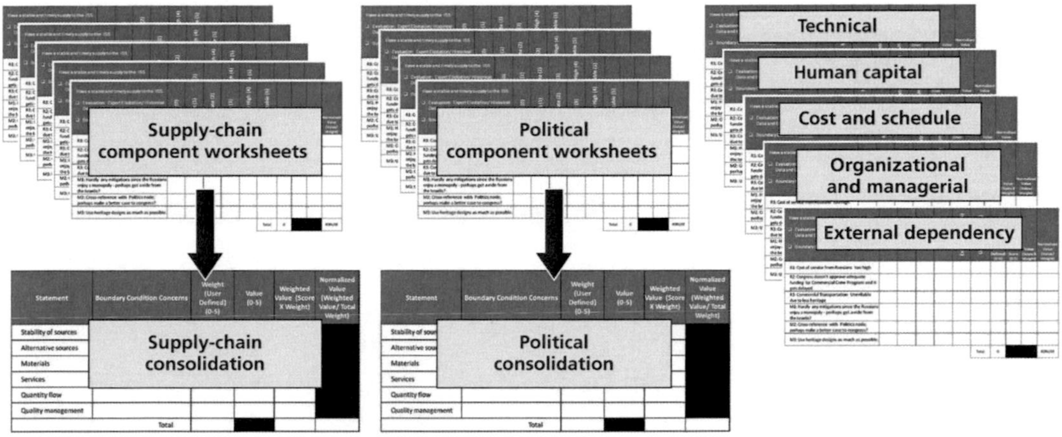

RAND RR1537-5.1

assessments for the technical, human capital, cost and schedule, organizational and managerial, and external dependency components (Figure 5.1).

In examining the supply-chain risk factor, we identified five components. We defined each of these, developed indicators, and identified evaluation methods and mitigation strategies for them. This work suggested quantitative evaluation measures of supply-chain risk would be appropriate. Figure 5.2 depicts the overarching defining of the supply-chain risk factor that we considered for the ISS transport evaluation. Many, but not all, of the risk factors and associated definitions were identical for our case studies.

Once we completed the initial definitional work for supply-chain risk, we developed the component risk worksheets. Figure 5.3 provides the risk factor component evaluation for stability-of-sources component.

For this risk factor component, we include a short definition along with the evaluation methodology and any boundary conditions that might exist. For the assessment of the stability-of-sources component, we determined expert elicitation, historical data, and other evidence to be appropriate. We found the boundary conditions vary by commodity and noted this. Such a finding complicates the analysis, but is vital to understanding the component and ultimately the outcome of the analysis.

We identified three components to the stability-of-sources risk: (1) Russia markedly raises the cost of service, (2) Congress does not approve adequate funding and thereby delays a commercial transportation program, and (3) commercial transportation is unreliable because it is less reliant on proven legacy systems. We also identified ways to mitigate each of these risks: (1) Find or subsidize an alternative supplier, (2) make a better case to Congress for supporting this effort, and (3) use legacy designs as much as possible.

Figure 5.2
Shuttle Cancellation/ISS Transport Initial Risk Factor Worksheet (Notional)

Supply Chain	Indicators	Evaluation Methods	Mitigation
Stability of sources for components and equipment			
Ability of commercial sector to provide stable supplies to the ISS in a timely manner	• Cost/schedule delays • Number of vendors • Performance • Reliability • Number of sources • Acceptable levels	Critical Path Methods; Qualitative	Accelerating readiness of supplies by investing more funds earlier; diversification of commercial partners; deferring retirement of Space Shuttle program
Availability of alternative sources (domestic and foreign) for components and equipment			
Alternative supply-chain sources and dependence on foreign countries	Alternative supply-chain sources available during each time period	Number of alternative supply-chain units required over time	Same as above
Availability of materials			
Availability of necessary materials	Units of material	Number of units available	Divide among options identified above
Availability of services			
Commercial sector to provide necessary services to the ISS	Units of services required per unit time	Number of units available	Divide among options identified above
Quality management			
Quality of the product delivered	Comparison to baseline quality	Quality assessment by subject matter	Rigorous quality management programs

RAND *RR1537-5.2*

There may not always be a one-to-one match between risk and mitigation. In some cases, there may be multiple mitigation strategies for a single risk. Alternatively, no mitigation may be possible.

Having identified the risk indicators, we established parameters corresponding to each risk indicator and mitigation value (i.e., low to unacceptable). For example, we assessed risk indicator 1 (R1) to be "low" if the actual cost was equal to the expected cost, and "some" if there was 10-percent cost growth. We repeated this process for all elements in the table. Because the risk of R1 was assessed as 25-percent growth being too high, it was given a score of 4. The corresponding mitigation M1 was deemed to reduce the risk to a score of 3.

Once we established thresholds for each risk and mitigation, we established weights for them. As noted, we recommend initially setting all weights to "1" and then adjusting as deemed necessary for the seriousness of the component. As a general rule, the risk indicators and their corresponding mitigations should have the same weight.

Figure 5.3
Shuttle Cancellation/ISS Transport Stability-of-Sources Risk Factor Component Worksheet (Notional)

Have a stable and timely supply to the ISS / Evaluation: Expert Elicitation Historical Data, Other Evidence / Boundary Conditions: each type of supply has different boundary conditions.	Low (0)	Some (1)	Moderate (2)	High (3)	Extremely High (4)	Unacceptable (5)	Weight (User Defined) (0–5)	Score (0–5)	Value (Score X Weight)	Normalized Value (Value/Total Weight)	Normalized Mitigated Value (Value/Total Weight)
R1: Russia markedly raises the cost of service.	Expected cost	10% too high	20% too high	30% too high	40% too high	50% too high	4	4	16	1.33	
R2: Congress does not approve adequate funding for Commercial Crew Program and it gets delayed.	Expected schedule	6-month delay	1-year delay	18-month delay	2-year delay	>2-year delay	5	2	6	0.50	
R3: Commercial Transportation unreliable due to less heritage.	Same reliability	0.1% less reliable	0.2% less reliable	0.3% less reliable	0.4% less reliable	>0.5% less reliable	5	2	10	0.83	
M1: Find alternative supplier, even if with subsidy.	Expected cost	10% too high	15% too high	20% too high	25% too high	30% too high	4	3	12		1.00
M2: Cross-reference with politics node; perhaps make a better case to Congress?	Expected schedule	6-month delay	1-year delay	18-month delay	2-year delay	>2-year delay	3	1	3		0.25
M3: Use heritage designs as much as possible.	Same reliability	0.1% less reliable	0.2% less reliable	0.3% less reliable	0.4% less reliable	>0.5% less reliable	5	1	5		0.42
						Total				2.67	1.67

NOTIONAL DATA ONLY

Using the evaluation methods identified for the risk factor component, we analyzed the individual risk indicators and mitigation strategies, relying on internal expertise and literature searches to arrive at the score. For an actual analysis, we would have employed NASA experts and more-formal evaluation procedures.

We then summed the scores, finding the unmitigated supply-chain risk score to be 2.67 and the mitigated one to be 1.67. This means that the unmitigated risk is in the "moderate" to "high" range, but that mitigation measures can reduce it to the "some" to "moderate" range. By tracking the costs associated with each mitigation strategy, we can calculate total costs for implementing some or all of the mitigations. (We did not do so for this case study because costs were not available.)

We repeated this process for the other four risk factor components and provide the results in Figure 5.4. Note that the supply-chain normalized risk factor value for the shuttle cancellation/ISS transport decision was assessed to be 2.82 or between "moderate" and "high." We could have conducted this same analysis using the mitigated risk values to arrive at a mitigated supply-chain risk assessment.

As noted, we conducted a detailed analysis for the political risk factor, while we used an abbreviated method for the other five risk factors that estimated risk directly by drawing holistically on a body of expertise. We did not include analysis of mitigation strategies in the abbreviated method.

Figure 5.5 provides the cost and schedule risk factor evaluation for the ISS transport decision. Because we only conducted an abbreviated analysis of this risk, we did not include details associated with boundary conditions and evaluation methods. For this particular factor, we calculated a score of 3.95, approaching the threshold of "extremely high" risk.

After evaluating all risk factors for the ISS transport decision, we plotted them on a radar chart. Figure 5.6 shows the overall scores for each risk factor, the risk thresholds, and the radar chart. This shows that cost and schedule risk poses the greatest risk to this decision among the risks we analyzed.

Our methodology provides a useful graphical depiction of the risks associated with the ISS transport decision. The chart in Figure 5.6 only provides unmitigated risk scores and therefore only a single curve, but it could accommodate a curve showing mitigated risk scores, thereby allowing comparison between them. Still other options could be considered and plotted, as we will show in the Cislunar Habitat case study.

The graphical analysis reflects only one part of the overall analysis. As noted earlier, the final graphical output is less important than the process of filling out the detailed worksheets; evaluating the risk indicators; determining the weights; and gaining an appreciation for the sensitivities for each risk factor and its components, indicators, and mitigation strategies. All this input provides a rich body of information and analysis that can contribute to interpretation of the results.

Figure 5.4
Shuttle Cancellation/ISS Transport Supply-Chain Risk Factor Component Evaluations (Notional)

Statement	Boundary Condition Concerns	Weight (User Defined) (0-5)	Value (0-5)	Weighted Value (Score X Weight)	Normalized Value (Weighted Value/ Total Weight)
Stability of sources	The point at which there is a lack of confidence in the stability of sources.	2	2.67	5.33	
Alternative sources	The alternative sources are unavailable or unreliable.	1		2.92	
Materials	Unavailability of material, or inability to transport them appropriately.	1	2.75	2.75	
Services	Unavailability of services available.	1	3.54	3.54	
Quality management	Inadequate quality of materials, services, or alternative sources.	1	2.40	2.40	
	Total	6		16.94	2.82

NOTIONAL DATA ONLY

Figure 5.5
Shuttle Cancellation/ISS Transport Cost and Schedule Evaluation (Alternative #1) (Notional)

Statement	Low (0)	Some (1)	Moderate (2)	High (3)	Extremely High (4)	Unacceptable (5)	Score (0-5)	Weight (User Defined) (0-5)	Value (Score X Weight)	Normalized Value (Value/Weight)
Cost for maintaining supply chains	Expected cost	10% too high	15% too high	20% too high	25% too high	30% too high	5	5	25	1.25
Schedule for maintaining supply chains	Expected schedule	1-week delay	2-week delay	3-week delay	4-week delay	>5-week delay	4	5	20	1
Budget stability	Expected stability	5% unstable	10% unstable	20% unstable	30% unstable	> 35% unstable	4	4	16	0.8
Program reliance on set cost and schedule	Program adjustable	25% adjustable	20% adjustable	10% adjustable	5% adjustable	<5% adjustable	3	3	9	0.45
Insource/outsource trade-offs in cost and flexibility	Trade-offs feasible	25% trade-off feasible	20% trade-off feasible	10% trade-off feasible	5% trade-off feasible	< 5% trade-off feasible	3	3	9	0.45
						Total		20		3.95

RAND RR1537-5.5

Figure 5.6
Shuttle Cancellation/ISS Transport Decision Risk Assessment (Notional)

Element	Risk Level
Political	2.77
Supply chain	2.82
Cost and schedule	3.95
Human capital	3.50
Organizational/managerial	2.10
External dependency	3.36
Technical	2.19

Level	
Unaccepta	5
Extremely High	4
High	3
Moderate	2
Some	1
Low	0

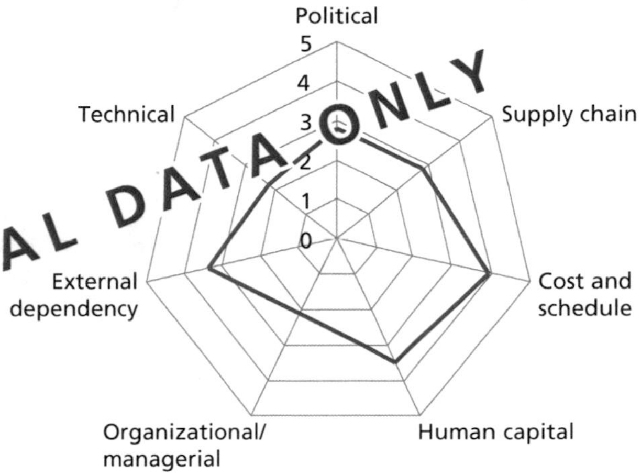

Cislunar Habitat

For our Cislunar Habitat case study, we conducted detailed risk factor assessments for the external dependency and organizational and managerial components, and abbreviated assessments for the technical, human capital, cost and schedule, political, and supply-chain components (Figure 5.7).

We used the process previously detailed for preparing the worksheets, establishing threshold values, and assessing the risk—with one notable exception. Because this issue is pending, we gave separate attention to several discrete paths under consideration. This allowed us to demonstrate the ease with which decisionmakers can visually gain insights into the differences of options under consideration.

The overarching case study examined the future inflection point at which a decision must be made to de-orbit ISS and transition to a Cislunar Habitat. Because a specific path has not been fully implemented, there still exist several likely scenarios, each with different risk factors.

The scenarios we considered are (1) a significant international partnership, (2) public-private partnerships, and (3) NASA-driven. The first option would have a Cislunar Habitat similar to the ISS: owned and operated by multiple countries. In the second option, different pieces of the architecture would be built by a combination of

Figure 5.7
Risk Assessment Worksheet for Cislunar Habitat

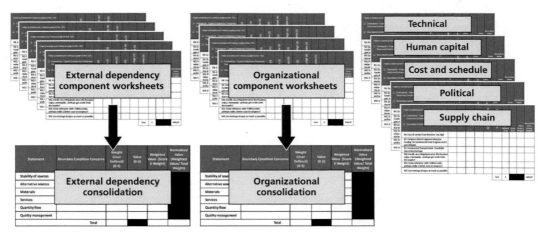

private and public space corporations. In the third option, an internal effort would use the Orion spacecraft and build toward human Mars missions.

The final results of the Cislunar Habitat analysis are depicted in Table 5.1 and Figure 5.8. As with the ISS transport analysis, the values reflect the unmitigated risks only. (The worksheets for the three options are provided online in the supplemental notional data for Appendix C.) The radar chart allows us to quickly deduce that the NASA-driven option poses the least risk, while the international cooperation option poses the greatest risk. Some of these observations are driven by specific types of risk. Two risk factors, cost and schedule and human capital, are equivalent for all three options. Supply-chain risk is slightly less for the international option, but the other four risk factors are less for the NASA-driven option, leading to our overall finding. Most noteworthy is the nearly unacceptable political risk for the international coopera-

Table 5.1
Cislunar Habitat Risk Assessment (Notional)

Risks	International Cooperation	Public-Private Partnerships	NASA-Driven
Political	4.67	3.67	3.00
Cost and schedule	3.90	3.90	3.90
Organizational and managerial	3.90	3.60	2.60
External dependency	3.90	3.32	2.00
Supply chain	2.94	3.70	3.53
Human capital	2.91	2.91	2.91
Technical	3.80	3.60	3.00

Figure 5.8
Cislunar Habitat Risk Assessment (Notional)

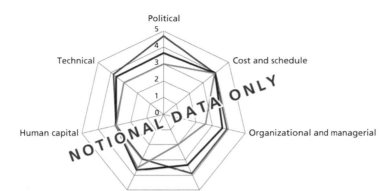

RAND *RR1537-5.8*

tion option. We caution that these values may change with a broader view from NASA experts, as well as from national, international, and public-private space entities.

Thoughts on the Case Studies

The case study analyses help validate the methodology. The normalization process provides a straightforward way to assess relative risks associated with each risk factor and the indicators and mitigations for each component. As such, it allows for easy assessment of which risk factors have the greatest or least amount of risk.

The discussions that accompany the completion of the component worksheets are the most important aspect of the methodology. As noted, the final graphical output reflected in the radar chart is less important than the process of filling out the detailed worksheets, conducting the evaluations on the risk indicators, determining the weights, and gaining an appreciation for the sensitivities for each risk factor and its components, risk indicators, and mitigation strategies. All this input provides a rich body of information and analysis that can contribute to interpretation of the results. Such results should be considered only in comparison, and not as absolute indicators of risk.

CHAPTER SIX

Findings and Conclusions

In this chapter, we summarize our accomplishments, our thoughts on the methodology we presented, and recommendations for future work.

Accomplishments

This research has achieved several noteworthy accomplishments. First, we described a methodology for assessing NASA-level risk, incorporating a variety of factors to be considered in such an assessment. We demonstrated that critical decisions could be identified and analyzed using this methodology. We also demonstrated that the methodology could work with various levels of fidelity and could delve into deeper layers when more-specific data are available. The methodology can also support sensitivity analysis.

Second, we examined the NASA system and developed a set of factors that completely define its risks. This provides a structured way to consider NASA-level risk, regardless of the issue under consideration. While these elements—the risk factors, risk factor components, and their associated risk indicators and metrics—are not independent, they represent a fairly complete articulation of the elements to be considered in a complex organizational risk assessment. An analyst can address dependencies by choosing the risk indicators and thresholds for normalization.

Third, our methodology provides a comprehensive approach for guiding staff and decisionmakers through a structured and repeatable process for assessing risk. It allows for deliberating on various risks involved in a decision and coming to a consensus about the likelihood and consequences of each risk. Through this process, NASA will be able to examine a wide range of complex, multidisciplinary issues.

Finally, we validated the methodology through case studies of a past and a future NASA decision. This demonstrated that the methodology could be employed in a retrospective and prospective manner.

Thoughts on the Methodology

While the methodology is robust and highly adaptable for a variety of cases, questions, and issues that NASA might face, it can require tailoring and would benefit from other resources. We discuss these below.

Overall Conclusions

Our experience indicates that the use of a structured methodology in conducting a risk-informed decision assessment far exceeds the benefits of the final calculations that result. Stated more directly, the process is far more important than the result.

The insights gained from employing the methodology and having to examine boundary conditions and thresholds force important understandings regarding the risk sensitivity for the issue under consideration. Consideration of not only the risks but also the mitigations at an early stage in the analysis prompts an important structured thinking process that will allow for clearer and more thoughtful decisions. While employing the methodology may not change decisions, gaining better understanding of key sensitivities will undoubtedly be an important outcome.

The use of a normalization process allows for comparing disparate risk issues that contribute to important decisions. Without such a process as part of the methodology, decisionmakers would be presented with a number of risk factors using different scales—and likely different grading and weighting—with little ability to understand either how risks compare with each other or the overall risk burden associated with a decision.

The normalization process also allows for determining relative component risks and even risk indicators and mitigation strategies. However, the outcomes of the normalization process should only be considered in a relative sense and not used for absolute comparisons in an absolute sense.[1] In other words, one may use our methodology to conclude that Option A is better than Option B, but not to conclude that Option A is 5 percent better than Option B.

Robustness of the Methodology

The methodology allows for tailoring to assess a wide variety of risk factors, risk components, risk indicators, and mitigation strategies, as well as to assess unmitigated or mitigated risk assessments. Further, in examining the mitigated risk associated with a particular decision, one can also attach costs (including time, personnel, and dollars) to the mitigation strategies and thereby develop a cost-informed set of mitigation strategies. The methodology also allows for comparing various options, as was demonstrated in the Cislunar Habitat case study with three options.

---

[1]   In the classical risk assessment format, when risks are represented by the likelihood of some outcome occurring (e.g., 10 percent of five fatalities), absolute values are indeed very important. Once these risk values have been normalized to a common ordinal scale (e.g., 1 to 5), the values become less descriptive and therefore lack the actionable aspect of their classical counterpart.

Requiring the analysis of each of the risk factors, components, indicators, and mitigations to begin with identification of the boundaries involved provides a clear articulation of the limits, acceptable and unacceptable, involved in the analysis. This development process allows senior leaders to articulate boundary conditions upon which a risk is no longer acceptable, which can guide the analysis.

The structured approach to conducting risk assessments also contributes to the understanding of the interdependencies associated with an issue. While eliminating all interdependencies in an issue might be desirable, doing so is unrealistic for the types of complex, multidisciplinary issues that NASA faces regularly. While one cannot eliminate these interdependencies, it is possible to define, understand, and assess how various relationships among the risk factors, components, indicators, and mitigations interact. Such reflection will provide insights regarding how best to manage and mitigate any of the negative effects of such interdependencies.

Some risk factors and indicators could be eliminated for certain issues. However, such an adaptation to the methodology should only be undertaken after careful deliberation.

Each Assessment Case Requires Tailoring to the Problem in Question

Each discrete assessment requires that the methodology being changed to reflect the particulars of the issue under consideration. In our case studies, these updates included the determination of new threshold values, weights, and scores for each of the risk indicators and mitigation strategies. These did, however, remain constant in the options we considered for the Cislunar Habitat case study, with only the scores updated to reflect the relative risks associated with each option.

Improving the Quality of the Analysis Requires Resources

The quality of the risk assessment is directly related to the quality of the inputs for the assessment. Access to actual data, subordinate risk calculations (e.g., technical assessments of components within programs), and expert elicitation would improve the risk assessments. Employing more-rigorous methods—such as probabilistic risk assessments, EVM, and decision trees—would also improve the ability to assess risk values.

Extensive resources may be required to develop the inputs for this methodology but once inputs are in hand, the only significant remaining task is to elicit weights from decisionmakers. The rest of the analysis could be automated with Excel worksheets. Inputs for each risk factor may be obtained from different experts in each area. Therefore, while resource-intensive, much of the work may occur concurrently.

Separating the Building of the Methodology from the Decisionmaking

Separating the experts from the decisionmakers is imperative for assuring the methodology produces unbiased results. Experts are essential in developing the risk factor component charts; decisionmakers are essential in developing weights. Decisionmakers should approach the methodology without being encumbered by the discussions associated with completing the chart and with the opportunity to synthesize the material and think at a higher level of abstraction about the results. That is, there should be an independent role for the decisionmaker, reflecting a broader pattern of thinking on a risk-informed decision issue.

Way Forward

The steps developed in the methodology provide a structured way to consider a risk-informed decision. While we were able to conduct abbreviated case studies validating it, conducting a more robust analysis for a future NASA decision is important. Given the work we have begun on it, applying this model to a more expanded analysis of the pending Cislunar Habitat decision may be most appropriate. Such work would provide further validation of the methodology and additional insights into the risk (and options) associated with this decision.

List of Potential Cases for Consideration

Joining Forces with the Russians for Building a Space Station

- **Time frame:** Decision made 1993, first module launched 1999, program funded through 2024
- **Description:** Throughout the 1980s, NASA was working on designing a large space station. With cost estimates increasing and designs becoming less and less ambitious, this effort was at risk of cancellation, but was invigorated when the possibility of collaboration with the Soviets/Russians became real in the early 1990s. Russia had much more experience with designing, building, and operating space stations at that time that the United States could tap into. A commitment to collaborate also meant that Congress would be less likely to cancel the program. Finally, this was seen as a way to keep Russian engineers and scientists gainfully employed, which otherwise would have posed a proliferation risk for strategic technologies. On the other hand, integrating Russian and American designs, processes, and procedures incurred technical, schedule, and financial risk and cost, and the United States now depended on Russian funding and time lines for some key contributions.

Cancellation of the Space Shuttle Program and Commercialization of Transport to the ISS

- **Time frame:** 2011
- **Description:** The decision was to commercialize transport to the ISS and focus NASA efforts on building the infrastructure for human missions to Mars. One key consideration is that the knowledge for building and operating space transportation systems has been transferred from one generation of NASA engineers to the next and the commercial sector lacks in this area.

Decision for Orion Capsule Splashdown and Navy Recovery

- **Time frame:** 2000s
- **Description:** The Orion capsule re-entry and landing will be a change from what NASA has known since 1981. It will harken back to the days of the early capsules that reentered the atmosphere and landed in the ocean. NASA will rely on the Navy for spacecraft recovery.

Mars Polar Lander Launched in January 1999

- **Time frame:** January 1999. Mission failed during Mars landing.
- **Description:** The decision was to deliver a lander to the surface of Mars for approximately one-half of the cost of Mars Pathfinder, which had been done for significantly less than earlier planetary missions. The following mandates were considered in order to meet the cost constraints:
  - Use off-the-shelf hardware components and inherited designs to the maximum extent possible.
  - Use analysis and modeling as an acceptable lower-cost approach to system test and validation.
  - Limit changes to those required to correct known problems; resist changes that do not manifestly contribute to mission success.

Developing Nuclear Power for Space Exploration

- **Time frame:** 2016–2030
- **Description:** Effective human exploration of the solar system will require nuclear fission–based power generators for surface bases. While reactor prototypes were built and tested decades ago (and a small-scale reactor provided power to the main U.S. Antarctic base in the 1960s), not much development has happened since because of political concerns and the lack of funded exploration missions. However, with human exploration beyond low-Earth orbit back on the national agenda, now is the time to examine the potential cost, benefits, and (technical and nontechnical) risks involved in restarting space nuclear reactor development. This analysis could be expanded to also look at in-space nuclear propulsion, in addition to surface power.

Designing Mars Surface Habitats

- **Time frame:** Next 25 years
- **Description:** Several options are being considered for a Mars surface habitat: underground base, 3-D printed housing, etc. What are the risks and benefits of each? What about the infrastructure used for human Mars exploration? Space helicopters? Cubesats?

Cislunar Space Station

- **Time frame:** 2020–2030
- **Description:** Follow-on to the ISS, the CLSS will be used as the next-generation outpost near Earth for technology demonstration and habitability in space. The goals are to improve on the lessons learned from ISS to improve life support, crew health, extravehicular activity capabilities, radiation shielding, and other areas. Much of the mission will be to validate processes, procedures, technology, and operations that would be used on a manned mission to Mars. The CLSS will give astronauts easier access to the moon, where testing can occur on a foreign astrological body close to Earth prior to committing to a Mars mission where recovery to Earth is measured in months instead of days.

Manned Missions to Mars

- **Time frame:** 2020–2040
- **Description:** Manned missions to Mars will be complex and expensive. One approach is for NASA to develop and execute the missions independently. Another is to use a coalition of international partners to share cost and technology development, much like the ISS.

Russian Spacecraft Access to the ISS

- **Time frame:** Ongoing decision
- **Description:** With the retirement of the U.S. Space Shuttle program, NASA decided to rely on equipment and services from both the government of the Russian Federation as well as Russian aerospace companies. Agreements between NASA and these organizations are controlled by the executive and legislative branches of the U.S. government. In the last few years, the relationship between NASA and Roskomos, the Russian Space Agency, has been the subject of discus-

sion due to U.S. sanctions on Russia following Russia's actions in Crimea. NASA is dependent on Russian rockets and launch services for access to the ISS.

Mars Surface Power Source

- **Time frame:** Next 25 years
- **Description:** What type of power source should humans use on the surface of Mars? Nuclear? Solar? What are the pros and cons and risks that each of these options induce?

Supersonic Test Aircraft

- **Time frame:** 2016–2025
- **Description:** NASA will develop a supersonic demonstrator in which to show that sonic booms can be dampened by changes in aircraft design and flight trajectories. The goal is to lift the limits of overland supersonic flight. The NASA project will need to explore partnering arrangements for technology development and cost sharing.

Cancellation of the Space Shuttle Program and Commercialization of Transport to the ISS Case Study

Supply Chain (in detail)

- **Stability of sources for components and equipment**[1]
- **Availability of alternative sources (domestic and foreign) for components and equipment**
- **Availability of materials**
- **Availability of services**
- **Quality management**

Cost and Schedule

- Cost associated with maintaining access to needed resources
- Schedule associated with maintaining access to needed resources
- Budget stability
- Program reliance on set cost and schedule
- Insource vs. outsource trade-offs in cost and flexibility

Human Capital

- Technical expertise
- Availability of talent
- Age of talent
- Cost of talent
- Adaptable skill mix vs. adaptability to changing missions
- Training programs in place

---

[1] Items in bold type are explored further in the online supplemental file, Appendix B, Cancellation of Space Shuttle and Commercialization of Transport to the ISS supplemental notional data (www.rand.org/t/rr1537).

Organizational and Managerial

- Strength and interest of leadership
- Levels of management involved in work
- Number of locations involved in work
- Dispersed management of projects and funds
- Congressional backing of individual locations
- Cultural differences between locations

External Dependencies

- Partnership and funding approaches for stability
- Level of dependence
- Amount of funding
- Primary mission, strategy, and planning of dependent organization
- Stability and strategy of dependent organization
- Alternatives for dependence

Political (in detail)

- **Changes in direction from political leaders of the country**
- **Reliance on foreign entities**
- **Congressional restrictions**
- **Federal budget process**

Technical

- Documentation for key technology cost and schedule assumptions
- Payoff of technology
- Readiness levels
- Safety of the technology systems
- Reliance on technology of other programs

Cislunar Habitat Case Study

Supply Chain

- Stability of sources for components and equipment[1]
- Availability of alternative sources (domestic and foreign) for components and equipment
- Availability of materials
- Availability of services
- Quality management

Cost and Schedule

- Cost associated with maintaining access to needed resources
- Schedule associated with maintaining access to needed resources
- Budget stability
- Program reliance on set cost and schedule
- Insource vs. outsource trade-offs in cost and flexibility

Human Capital

- Technical expertise
- Availability of talent
- Age of talent
- Cost of talent
- Adaptable skill mix vs. adaptability to changing missions
- Training programs in place

---

[1] Items in bold type are explored further in the online supplemental file Appendix C, Cislunar Habitat supplemental notional data (www.rand.org/t/rr1537).

Organizational and Managerial (in detail)

- **Strength and interest of leadership**
- **Levels of management involved in work**
- **Number of locations involved in work**
- **Dispersed management of projects and funds**
- **Congressional backing of individual locations**
- **Cultural differences between locations**

External Dependencies (in detail)

- **Partnership and funding approaches for stability**
- **Level of dependence**
- **Amount of funding**
- **Primary mission, strategy, and planning of dependent organization**
- **Stability and strategy of dependent organization**
- **Alternatives for dependence**

Political

- Changes in direction from political leaders of the country
- Reliance on foreign entities
- Congressional restrictions
- Federal budget process

Technical

- Documentation for key technology cost and schedule assumptions
- Payoff of technology
- Readiness levels
- Safety of the technology systems
- Reliance on technology of other programs

Human Capital Risk Methodology

As discussed in the body of the report, NASA is composed of ten centers that are located around the country. NASA is somewhat unique in that the centers themselves were independent research laboratories prior to the creation of NASA, and therefore certain cultural aspects remain. While somewhat unique, there are some examples of managing similar risks which NASA could use. One such example is used by the Naval Sea Systems Command (NAVSEA). NAVSEA's methods are discussed here.

Human Capital Risk Methodology

Assessing the risk associated with human capital can make use of existing concepts. Individual competency or skills can be tracked and assessed annually. Performing this task will cover many of the metrics that were listed earlier.

An external example of managing human capital can be seen in the U.S. Navy's Warfare Centers (WFCs). NASA centers and WFCs have similarities in that they are field activities that contain the technical skill and equipment necessary for their parent organizations to perform missions. Each year the WFCs perform what is known as the Technical Capability Health Assessment (TCHA). The centers go through their technical capabilities (TCs) and rate them on a scale from "Improved" to "Unchanged" to "Declined." Based on the scoring, NAVSEA can decide where to invest in equipment, facilities, and people (Figure D.1).

Individuals at the WFCs are fitted into the TCs using a system called "knowledge areas." The WFCs rate each employee based on his or her background as it fits into the WFCs' TCs. Some WFCs allow employees to have knowledge areas that fit in multiple TCs; some only allow one knowledge area per person. In both cases, the WFCs are able to identify skill gaps, and understand how the workforce changes over time as TCs are added and removed from the different WFCs. Sometimes, when a TC is removed from a given center, employees are left with knowledge areas that do not fit the remaining TCs. In these cases, employees are given opportunities to move to the center with the TC, engage in a reeducation process that would allow them to fit into a remaining TC, or other tailored processes. Sometimes workers who remain with no TC are simply left

Figure D.1
Technical Capabilities Supply and Demand Scoring

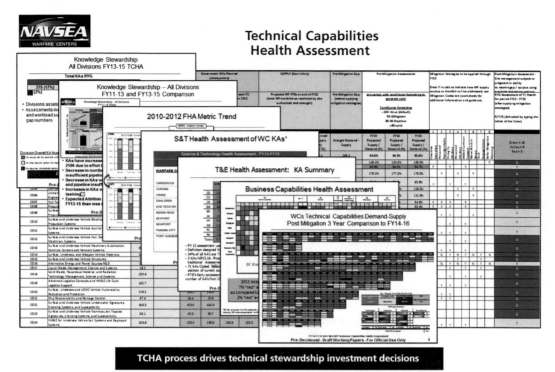

SOURCE: NAVSEA, email with author, January 27, 2016.
RAND RR1537-D.1

alone until their career ends, but this costs the Navy money as the employees no longer have specialized work that they can perform.

References

Anbari, Frank T., "Earned Value Project Management Method and Extensions," *Project Management Journal*, Vol. 34, No. 4, 2003, pp. 12–23.

Armagh Planetarium, "Skylab: Everything You Need to Know," *Astronotes* blog, May 1, 2013. As of April 20, 2016:
http://www.armaghplanet.com/blog/skylab-everything-you-need-to-know.html

Arrow, Kenneth Joseph, *Aspects of the Theory of Risk-Bearing*, Helsinki: Academic Bookstores, 1965.

Atwood, Corwin, Lauren Fleishman, Jon Johnson, Clifford Marks, Scott Newberry, and Robert Youngblood, *Technical Report on the Prioritization of Inspection Resources for Inspections, Tests, Analyses and Acceptance Criteria (ITAAC)*, Rockville, Md.: Information Systems Laboratories, Inc., submitted to U.S. Nuclear Regulatory Commission, 2005.

Ayyub, Bilal M., *Elicitation of Expert Opinions for Uncertainty and Risks*, Boca Raton, Fla.: CRC Press, 2001a.

———, *A Practical Guide on Conducting Expert-Opinion Elicitation of Probabilities and Consequences for Corps Facilities*, Alexandria, Va.: U.S. Army Corps of Engineers, IWR Report 01-R-01, January 2001b. As of April 20, 2016:
http://www.iwr.usace.army.mil/Portals/70/docs/iwrreports/01-R-01.pdf

Barron, F. Hutton, and Bruce E. Barrett, "Decision Quality Using Ranked Attribute Weights," *Management Science*, Vol. 42, No. 11, 1996, pp. 1515–1523.

Bond, Craig A., Lauren A. Mayer, Michael E. McMahon, James G. Kallimani, and Ricardo Sanchez, *Developing a Methodology for Risk-Informed Trade-Space Analysis in Acquisition*, Santa Monica, Calif.: RAND Corporation, RR-701-A, 2015. As of May 31, 2016:
http://www.rand.org/pubs/research_reports/RR701.html

Boyer, Roger L., and Teri Hamlin, *Use of Probabilistic Risk Assessment in Shuttle Decision Making Process*, Houston, Tex.: NASA Johnson Space Center, 2011.

Broad, William J., "For Parts, NASA Boldly Goes . . . on eBay," *New York Times*, May 12, 2002. As of April 20, 2016:
http://www.nytimes.com/2002/05/12/us/for-parts-nasa-boldly-goes-on-ebay.html

Budescu, David V., and Thomas S. Wallsten, "Consistency in Interpretation of Probabilistic Phrases," *Organizational Behavior and Human Decision Processes*, Vol. 36, No. 3, 1985, pp. 391–405.

Carney, Jordain, "Senators Battle over Russian Rocket Engines," *The Hill*, December 16, 2015. As of April 20, 2016:
http://thehill.com/blogs/floor-action/senate/263525-senators-battle-over-russian-rocket-engines

Clark, Charles S., "OMB Prepares to Ratchet Up Enterprise Risk Management," *Government Executive*, February 29, 2016. As of June 6, 2016:
http://www.govexec.com/management/2016/02/
omb-prepares-ratchet-enterprise-risk-management/126297/

Clemen, Robert T., and Terence Reilly, *Making Hard Decisions with Decision Tools*, Mason, Ohio: South-Western Cengage Learning, 2001.

Clemen, Robert T., and Robert L. Winkler, "Combining Probability Distributions from Experts in Risk Analysis," *Risk Analysis*, Vol. 19, No. 2, 1999, pp. 187–203. As of April 20, 2016:
http://rd.springer.com/article/10.1023%2FA%3A1006917509560?LI=true

Cox, Louis Anthony Tony, "What's Wrong with Risk Matrices?" *Risk Analysis*, Vol. 28, No. 2, 2008, pp. 497–512.

Curtright, Aimee E., M. Granger Morgan, and David W. Keith, "Expert Assessments of Future Photovoltaic Technologies," *Environmental Science & Technology*, Vol. 42, No. 24, 2008, pp. 9031–9038.

Davis, Paul K., and Paul Dreyer, *RAND's Portfolio Analysis Tool (PAT): Theory, Methods, and Reference Manual*, Santa Monica, Calif.: RAND Corporation, TR-756-OSD, 2009. As of May 31, 2016:
http://www.rand.org/pubs/technical_reports/TR756.html

Davis, Paul K., Russell D. Shaver, and Justin Beck, *Portfolio-Analysis Methods for Assessing Capability Options*, Santa Monica, Calif.: RAND Corporation, MG-662-OSD, 2008. As of May 31, 2016:
http://www.rand.org/pubs/monographs/MG662.html

Dawes, Robyn M., "The Robust Beauty of Improper Linear Models in Decision Making," *American Psychologist*, Vol. 34, No. 7, 1979, p. 571.

Dawes, Robyn M., and B. Corrigan, "Linear Models in Decision Making," *Psychological Bulletin*, Vol. 81, 1974, pp. 182–196.

Dezfuli, Homayoon, Michael Stamatelatos, Gaspare Maggio, Christopher Everett, Robert Youngblood, et al., *NASA Risk-Informed Decision Making Handbook*, Washington, D.C.: Office of Safety and Mission Assurance, NASA Headquarters, NASA/SP-2010-576, Version 1.0, April 2010.

Edwards, Ward, and F. Hutton Barron, "SMARTS and SMARTER: Improved Simple Methods for Multiattribute Utility Measurement," *Organizational Behavior and Human Decision Processes*, Vol. 60, No. 3, 1994, pp. 306–325.

Federal Aviation Administration, *Risk Management Handbook*, Washington, D.C.: U.S. Department of Transportation, FAA-H-8083-2, 2009. As of April 20, 2016:
http://www.faa.gov/regulations_policies/handbooks_manuals/aviation/media/faa-h-8083-2.pdf

Ferdous, Nazneen, Lauren A. Mayer, George Hart, Peter Burge, and Colin Smith, *Application of Fair Division, Data Envelopment Analysis, and Conjoint Analysis Techniques to Funding Decisions at the Program and Project/Activity Level*, American Association of State Highway and Transportation Officials, Standing Committee on Planning, December 2014. As of April 20, 2016:
http://onlinepubs.trb.org/onlinepubs/nchrp/docs/NCHRP08-36(115)_FR.pdf

Fischhoff, Baruch, Paul Slovic, Sarah Lichtenstein, Stephen Read, and Barbara Combs, "How Safe Is Safe Enough? A Psychometric Study of Attitudes Towards Technological Risks and Benefits," *Policy Sciences*, Vol. 9, No. 2, 1978, pp. 127–152.

Fleming, Quentin W., and Joel M. Koppelman, "Earned Value Management: Mitigating the Risks Associated with Construction Projects," *Program Manager*, Vol. 31, No. 2, 2002, pp. 90–95.

Frank, Michael V., "Choosing Among Safety Improvement Strategies: A Discussion with Example of Risk Assessment and Multi-Criteria Decision Approaches for NASA," *Reliability Engineering & System Safety*, Vol. 49, No. 3, 1995, pp. 311–324.

Galway, Lionel A., *Subjective Probability Distribution Elicitation in Cost Risk Analysis: A Review*, Santa Monica, Calif.: RAND Corporation, TR-410-AF, 2007. As of June 1, 2016: http://www.rand.org/pubs/technical_reports/TR410.html

Garvey, Paul R., and Zachary F. Lansdowne, "Risk Matrix: An Approach for Identifying, Assessing, and Ranking Program Risks," *Air Force Journal of Logistics*, Vol. 22, No. 1, 1998, pp. 18–21.

Hanes, Elizabeth, "The Day Skylab Crashed to Earth: Facts About the First U.S. Space Station's Re-Entry," History.com, July 11, 2012. As of April 20, 2016: http://www.history.com/news/the-day-skylab-crashed-to-earth-facts-about-the-first-u-s-space-stations-re-entry

Hastie, Reid, and Robyn M. Dawes, *Rational Choice in an Uncertain World: The Psychology of Judgment and Decision Making*, 2nd ed., Thousand Oaks, Calif.: SAGE Publications, 2010.

Herridge, Linda, "Orion Crew Module Recovered from Pacific Ocean After First Flight Test," NASA John F. Kennedy Space Center, December 10, 2014. Photo credited to U.S. Navy. As of April 20, 2016: http://www.nasa.gov/content/nasa-us-navy-recover-orion-from-pacific-ocean

Holloway, Charles A., *Decision Making Under Uncertainty: Models and Choices*, Englewood Cliffs, N.J.: Prentice-Hall, 1979.

Homeland Security Institute, *Department of Homeland Security Science and Technology Readiness Level Calculator* (Ver 1.1) *Final Report and User's Manual*, Arlington, Va.: Homeland Security Studies and Analysis Institute, September 30, 2009. As of April 20, 2016: http://www.homelandsecurity.org/docs/reports/DHS_ST_RL_Calculator_report20091020.pdf

International Organization for Standardization, *ISO 31000—Risk Management*, 2009. As of April 20, 2016: http://www.iso.org/iso/home/standards/iso31000.htm

International Risk Governance Council, *An Introduction to the IRGC Risk Governance Framework*, Geneva, 2008. As of April 20, 2016: http://www.irgc.org/IMG/pdf/An_introduction_to_the_IRGC_Risk_Governance_Framework.pdf

Kahneman, Daniel, Paul Slovic, and Amos Tversky, eds., *Judgment Under Uncertainty: Heuristics and Biases*, Cambridge, UK: Cambridge University Press, 1982.

Kaplan, Robert S., and David P. Norton, "Using the Balanced Scorecard as a Strategic Management System," *Harvard Business Review*, 1996.

Keeney, Ralph L., "Multiplicative Utility Functions," *Operations Research*, Vol. 22, No. 1, 1974, pp. 22–34.

———, "Utility Functions for Equity and Public Risk," *Management Science*, Vol. 26, No. 4, 1980, pp. 345–353.

———, *Value-Focused Thinking: A Path to Creative Decisionmaking*, Boston, Mass.: Harvard University Press, 2009.

Keeney, Ralph L., and Howard Raiffa, *Decision with Multiple Objectives*, New York: Wiley, 1976.

Kelly, Kathryn E., and N. C. Cardon, "The Myth of 10^{-6} as a Definition of Acceptable Risk," 84th Annual Meeting and Exhibition of the Air and Waste Management Association, Vancouver, British Columbia, 1991.

Kremer, Ken, "ISS, NASA and US National Security Dependent on Russian and Ukrainian Rocketry Amidst Crimean Crisis," *Universe Today*, March 5, 2014. As of April 20, 2016: http://www.universetoday.com/110006/iss-nasa-and-us-national-security-dependent-on-russian-ukrainian-rocketry-amidst-crimean-crisis/

Mai, Thuy, ed., "Technology Readiness Level," NASA website, updated July 31, 2015. As of April 20, 2016: https://www.nasa.gov/directorates/heo/scan/engineering/technology/txt_accordion1.html

Mankins, John C., "Technology Readiness Assessments: A Retrospective," Acta Astronautica, Vol. 65, No. 9, 2009, pp. 1216–1223.

Markel, M. Wade, Bryan W. Hallmark, Peter Schirmer, Louay Constant, Jaime L. Hastings, Henry A. Leonard, Kristin J. Leuschner, Lauren A. Mayer, Caolionn O'Connell, Christina Panis, Jose Rodriguez, Lisa Saum-Manning, and Jonathan Welch, *A Preliminary Assessment of the Regionally Aligned Forces (RAF) Concept's Implications for Army Personnel Management*, Santa Monica, Calif.: RAND Corporation, RR-1065-A, 2015. As of May 31, 2016: http://www.rand.org/pubs/research_reports/RR1065.html

Massingham, Peter, "Knowledge Risk Management: A Framework," *Journal of Knowledge Management*, Vol. 14, No. 3, 2010, pp. 464–485. As of June 20, 2016: http://lpis.csd.auth.gr/mtpx/km/material/JKM-14-3c.pdf

Mohaghegh, Zahra, Reza Kazemi, and Ali Mosleh, "Incorporating Organizational Factors into Probabilistic Risk Assessment (PRA) of Complex Socio-Technical Systems: A Hybrid Technique Formalization," *Reliability Engineering & System Safety*, Vol. 94, No. 5, 2009, pp. 1000–1018.

Morgan, Jim, "Manufacturing Readiness Levels (MRLs) and Manufacturing Readiness Assessments (MRAs)," briefing, Air Force Research Lab, Manufacturing Technology Division, Wright-Patterson Air Force Base, Ohio, 2008. As of April 20, 2016: http://www.dtic.mil/cgi-bin/GetTRDoc?Location=U2&doc=GetTRDoc.pdf&AD=ADA510027

Morgan, M. Granger, Peter J. Adams, and David W. Keith, "Elicitation of Expert Judgments of Aerosol Forcing," *Climatic Change*, Vol. 75, No. 1–2, 2006, pp. 195–214.

Morgan, M. Granger and Max Henrion, *Uncertainty: A Guide to Dealing with Uncertainty in Quantitative Risk and Policy Analysis*, Cambridge, UK: Cambridge University Press, 1990.

NASA—*See* National Aeronautics and Space Administration.

National Aeronautics and Space Administration, "NASA Centers and Facilities," NASA Shared Services Center web page, undated. As of April 20, 2016: http://www.nasa.gov/about/sites/index.html

———, *NASA Human Capital 2015: A Guide to Building a Highly Engaged Workforce*, 2015a. As of April 20, 2016: http://nasapeople.nasa.gov/hcm/index_sbg.htm

———, "International Space Station," last updated November 24, 2015b. As of April 20, 2016: http://www.nasa.gov/mission_pages/station/expeditions/expedition01/index.html

National Research Council, "Appendix B: Review of Acceptable Cancer Risk Levels," in *Review of the Army's Technical Guides on Assessing and Managing Chemical Hazards to Deployed Personnel*, Washington, D.C.: National Academies Press, 2004, pp. 137–144. As of April 20, 2016: http://www.nap.edu/read/10974/chapter/9

Office of Audits, *NASA's Challenges to Meeting Cost, Schedule, and Performance Goals*, Office of the Inspector General, National Aeronautics and Space Administration, IG-12-021 (Assignment No. A-11-009-00), September 27, 2012. As of April 20, 2016:
https://oig.nasa.gov/audits/reports/FY12/IG-12-021.pdf

Office of the Secretary of Defense Manufacturing Technology Program in collaboration with The Joint Service/Industry MRL Working Group, *Manufacturing Readiness Level (MRL) Deskbook*, Version 2.0, Office of the Secretary of Defense, May 2011. As of April 20, 2016:
http://static1.squarespace.com/static/55ae48f4e4b0d98862c1d3c7/t/55df0e7ee4b05aef5ffb b5b2/1440681598134/MRL_Deskbook_V2.pdf

Osburg, Jan, Philip S. Anton, Frank Camm, Jeremy M. Eckhause, Jaime L. Hastings, Jakub Hlavka, James G. Kallimani, Thomas Light, Chad J. R. Ohlandt, Douglas Shontz, Abbie Tingstad, and Jia Xu, *Expanding Flight Research: Capabilities, Needs, and Management Options for NASA's Aeronautics Research Mission Directorate*, Santa Monica, Calif.: RAND Corporation, RR-1361-NASA, 2016. As of July 1, 2016:
http://www.rand.org/pubs/research_reports/RR1361.html

Paté-Cornell, E., and R. Dillon, "Probabilistic Risk Analysis for the NASA Space Shuttle: A Brief History and Current Work," *Reliability Engineering & System Safety*, Vol. 74, No. 3, December 2001, pp. 345–352.

Pratt, John W., "Risk Aversion in the Small and in the Large," *Econometrica: Journal of the Econometric Society*, Vol. 32, No. 1–2, 1964, pp. 122–136.

Program Executive Office Ships, "Keel Laid for Latest Addition to Multimission-Capable Amphibious Fleet," U.S. Navy website, July 18, 2009. As of April 20, 2016:
http://www.navy.mil/submit/display.asp?story_id=47036

Sadowitz, March, and John D. Graham, "A Survey of Residual Cancer Risks Permitted by Health, Safety and Environmental Policy," *Risk*, Vol. 6, 1995, p. 17.

Sauser, Brian J., Michael Long, Eric Forbes, and Suzanne E. McGrory, "Defining an Integration Readiness Level for Defense Acquisition," International Symposium of the International Council on Systems Engineering, Singapore, July 20–23, 2009.

Schoemaker, Paul J. H., and C. Carter Waid, "An Experimental Comparison of Different Approaches to Determining Weights in Additive Utility Models," *Management Science*, Vol. 28, No. 2, 1982, pp. 182–196.

Slovinac, Patricia, and Joan Deming, *Avionics Systems Laboratory/Building 16: Historical Documentation*, Washington, D.C.: NASA, JSC-CN-22284, June 2010.

Society for Risk Analysis, *SRA Glossary*, June 22, 2015. As of April 20, 2016:
http://www.sra.org/sites/default/files/pdf/SRA-glossary-approved22june2015-x.pdf

SRA—*See* Society for Risk Analysis.

Stillwell, William G., David A. Seaver, and Ward Edwards, "A Comparison of Weight Approximation Techniques in Multiattribute Utility Decision Making," *Organizational Behavior and Human Performance*, Vol. 28, No. 1, 1981, pp. 62–77.

Stone, Jeff, "Elon Musk: SpaceX 'Complacency' Contributed to Falcon 9 Crash, Falcon Heavy Rocket Debuts in 2016," *International Business Times*, July 21, 2015. As of April 20, 2016:
http://www.ibtimes.com/elon-musk-spacex-complacency-contributed-falcon-9-crash-falcon-heavy-rocket-debuts-2017809

Suciu, Peter, "End of Space Shuttle Program to Have Far Reaching Impact," CNBC, July 8, 2011. As of April 20, 2016:
http://www.cnbc.com/id/43469916

U.S. Department of Defense, *Risk Management Guide for DoD Acquisition*, 6th ed., Version 1.0, Washington, D.C., 2006. As of April 20, 2016:
http://www.dau.mil/publications/publicationsDocs/RMG%206Ed%20Aug06.pdf

U.S. Department of Homeland Security, *Risk Management Fundamentals: Homeland Security Risk Management Doctrine*, Washington, D.C., April 2011. As of April 20, 2016:
https://www.dhs.gov/xlibrary/assets/rma-risk-management-fundamentals.pdf

von Neumann, John, and Oskar Morgenstern, *Theory of Games and Economic Behavior*, Princeton, N.J.: Princeton University Press, 1944; 2nd ed., 1947; 3rd ed., 1953.

von Winterfeldt, Detlof, and Ward Edwards, *Decision Analysis and Behavioral Research*, New York: Cambridge University Press, 1986.

von Winterfeldt, Detlof, and Gregory W. Fischer, "Multiattribute Utility Theory: Models and Assessment Procedures," *Utility, Probability, and Human Decision Making*, Vol. 11, 1975, pp. 47–85.

Weber, P., G. Medina-Oliva, C. Simon, and B. Iung, "Overview on Bayesian Networks Applications for Dependability, Risk Analysis, and Maintenance Areas," *Engineering Applications of Artificial Intelligence*, Vol. 25, No. 4, June 2012, pp. 671–682.

Wainer, Howard, "Estimating Coefficients in Linear Models: It Don't Make No Nevermind," *Psychological Bulletin*, Vol. 83, No. 2, 1976, p. 213.